Exploring Mercury
The Iron Planet

Springer

London
Berlin
Heidelberg
New York
Hong Kong
Milan
Paris
Tokyo

Left: a rare evening alignment, 28 April, 2002. Richard Hill, Tuscon, Arizona photographed five planets that were alighned in the evening sky throughout April and May, 2002. Mercury is low, but clearly visible above the tree tops. Right: A month later, Robert McMillan, Tuscon, Arizona, photographed Mercury rising before the Sun. Mercury is the object (marked with a short line) in the centre-left of the picture where it is rising with the Hyades star cluster. Aldebaran is the bright star just over the tree tops to the right

Mercury makes about six appearances in the morning or evening twilight every year. Some appearances, or *apparitions*, are much better for viewing from the northern hemisphere than others which are better viewed from the southern hemisphere. The images above show Mercury viewed from Tuscon, Arizona.

Robert G. Strom and Ann L. Sprague

Exploring Mercury

The Iron Planet

Springer

Published in association with
Praxis Publishing
Chichester, UK

Professor Robert G. Strom
Professor Emeritus
Department of Planetary Sciences
University of Arizona
Tucson
Arizona
USA

Dr Ann L. Sprague
Lunar and Planetary Laboratory
University of Arizona
Tucson
Arizona
USA

SPRINGER–PRAXIS BOOKS IN ASTRONOMY AND SPACE SCIENCES
SUBJECT *ADVISORY EDITOR*: John Mason M.Sc., B.Sc., Ph.D.

ISBN 1-85233-731-1 Springer-Verlag Berlin Heidelberg New York

British Library Cataloguing in Publication Data
Strom, Robert G.
    Exploring Mercury : the iron planet. — (Springer-Praxis
    books in astromony and space sciences)
    1. Mercury (Planet)
    I. Title   II. Sprague, Anne L.
    523.4′1
    ISBN 1852337311

Library of Congress Cataloging-in-Publication Data
Strom, Robert G.
    Exploring Mercury : the iron planet / Robert G. Strom and Ann L. Sprague
        p. cm. — (Springer-Praxis books in astronomy and space sciences)
    Includes bibliographical references and index.
    ISBN 1-85233-731-1 (acid-free paper)
        1. Mercury (Planet)   I. Sprague, Ann L.   II. Title.   III. Series.
    QB611.S767 2003
    523.41–dc21   2003042781

Cover design: Jim Wilkie
Project Management: Originator Publishing Services, Gt. Yarmouth, Norfolk, UK

Printed on acid-free paper

Contents

Preface

NASA will not observe Mercury with the Hubble Space Telescope (HST) because it is too close to the Sun. If something should go wrong with pointing the HST at Mercury and it looked at the Sun it could ruin much of the instrumentation. There has only been one space mission to Mercury, the *Mariner 10* mission, 29 years ago, between 1974 and 1975. While the mission resulted in three flybys, each flyby saw the same side of the planet. Therefore, only 45% of the planet has been seen in detail. Fortunately, recent ground-based optical and radar observations have contributed important new information about Mercury that has reawakened a keen interest in learning more about this neglected planet. The recent imaging of the "unseen side", the discovery of possible ices at high latitudes, and the apparent absence of "lunar like" iron-rich basalts have added to Mercury's mystery. We have attempted to summarize all the exciting recent discoveries about Mercury and to place them in the context of planetary origins and history. We hope this book serves as a useful reference for future studies, and, particularly, for the upcoming missions to Mercury at the end of this decade.

This book is dedicated to all our friends, known and unknown,
who have recognized that detailed knowledge of Mercury and its formation is critical
to understanding the formation of our solar system and others.

Acknowledgements

The knowledge described in this book is the accumulated wisdom of many scientists accrued with the help of engineers and personnel in spacecraft control centers and telescope support facilities. It is impossible for us to explicitly name them all. But we recognize that without them we would not have enjoyed the privilege of writing this book. We accept the responsibility for any inevitable errors which lie herein and we hope and expect that future exploration, both ground-based and from space, will necessitate revisions and additions to this text. We greatly appreciate helpful reviews of material in this book by Mario Acuña, James Hays, Donald Hunten, John Lewis, Andrew Potter, Mark Robinson, and James Slavin. In addition, many pleasant and useful discussions with Josh Emery, Martha Leake, and Johan Warell contributed to the content and improvement of the manuscript. Maria Schuchardt, the data manager for the Lunar and Planetary Laboratory, Space Imagery Center provided cheerful and valuable support for processing many of the images that enhance this description of Mercury. We would also like to thank James McAdams and Ralph McNutt for providing material on the MESSENGER project.

Figures

Tables

Colour plates (between pages 114 and 115)

2.2 This illustration of *Mariner 10* shows the solar panels, long magnetometer boom, the sun shield, the high-gain antenna, and the science instruments. The pole-like object at the top is the omnidirectional, low-gain antenna.

2.3 At the moment the launch window opened, *Mariner 10* was sent on its way to Mercury on 3 November, 1973 at 12:45 a.m. Eastern Standard Time.

2.7 The first flyby of Mercury took place on the planet's night side. This illustration shows *Mariner 10*'s flightpath by Mercury and some of the image footprints and infrared traces taken on the planet's surface.

2.10 The second flyby on Mercury's day side permitted the acquisition of images of the southern hemisphere which joined the two sides seen on the first flyby (see Figure 2.12).

4.8 *Mariner 10* photomosaic of the incoming side together with an accurate artist's rendition of the size of Mercury's core compared to the silicate portion. The outer part of the core may still be in a liquid state (from Strom, 1987).

5.2 This concept of Mercury's magnetic field is based upon the *Mariner 10* data and subsequent modeling and analyses of those data. It is possible that this view will change with new observations made by particles and fields instrumentation on future missions to Mercury (courtesy of James Slavin, NASA Goddard Space Flight Center, Laboratory for Extraterrestrial Physics, Greenbelt, Maryland, USA).

6.2 Images of the sodium emission intensity in kilo-Rayleighs (KR) measured anti-sunward of Mercury at 0300 UT 26 May, 2001. Each square represents an observation with a 10×10 arcsec image slicer, and represents an area 5100 km. The emission intensity varies over four orders of magnitude, decreasing from about 6000 KR at the planet to about 0.6 KR at the extreme end of the tail. The T shape is not part of the shape of the Na cloud (Courtesy of Andrew Potter).

6.3 Images of Mercury's atmospheric Na emission show bright spots. Some scientists think these are spots where ions directed toward the surface by electric fields near Mercury have sputtered neutral Na off the surface. Other

6.4 (a) Radar image of spots A and B. (b) Diagram of geometry for slit spectroscopy measurements of Na enhancements over spots A and B (from Sprague *et al.*, 1997). (c) Na enhancements over spot K, A, and an unidentified region. (d) Na enhancements from spectroscopic measurements interpolated to make an "image" (from Sprague *et al.*, 1998). (e) Another case of Na enhancements over K, combined with B and A. (f) Spectroscopic measurements when the slit was placed over Caloris basin and an unidentified source. (g) Na enhancement over Caloris basin. (h) Na enhancements over the regions of bright albedo features at: 155, 65°N; 125, 0; and 105, 9°S longitude and latitude respectively. (i) Enhancement of Na spreading from ~43–73° longitude and ~10–60°S latitude

scientists believe the Na is bright at these locations because there are freshly exposed Na rich rocks and soils at those regions and that the Na is released when the regolith is heated. (Adapted from Potter and Morgan, 1990).

10.4 Paleogeologic maps of Mercury's incoming side seen by *Mariner 10* showing the distribution of craters and plains of various relative ages. The oldest craters (C_4-C_5) and plains (P_4-P_5) pre-date the hilly and lineated terrain (green) and are probably pre-Caloris impact age. The C_5 craters predate the P_4-P_5 plains. The youngest craters (C_1-C_2) and plains (P_1-P_5) post-date the hilly and lineated terrain (green) and are probably post-Caloris impact age. The P_5-P_3 plains are equivalent to intercrater plains and the P_2-P_1 plains are smooth plains (from Leake, 1981).

10.10 This false color photomosic by Mark Robinson shows sharp color boundaries that coincide with geologic boundaries. They probably represent differences in composition. The Kuiper/Muraski crater complex is indicated by "K". The "F" and "B" indicates plains units with sharp color boundaries. The "B"s are plains flooded craters. The largest is Lermontov (160 km diameter). The "blue" area at "D" may be material ejected from the subsurface by a fresh impact crater. See text for explanation (courtesy of Mark Robinson, Northwestern University).

13.1 This diagram shows the trajectory of the *MESSENGER* spacecraft as it flies by Venus and Mercury before Mercury orbit insertion on 5 April, 2009. The Deep Space Maneuvers (DSM) are engine burns to correct the trajectory of the spacecraft. The ΔV is the change in velocity required by that maneuver. At the bottom of the diagram is a schematic timeline of the mission up to Mercury orbit insertion with the number of spacecraft solar orbits in parentheses (courtesy of James McAdams, Applied Physics Laboratory, Laurel, MD).

13.3 Schematic illustration of the *MESSENGER* spacecraft showing positions of the science instruments and other features.

Information on CD (inside back cover)

This CD contains an extensive set of the highest resolution *Mariner 10* images of Mercury. There are several parts to this record. The folder labelled "Large Mosaics" contains the mosaics of the entire incoming and outgoing sides of the planet as viewed by *Mariner 10*. It also includes the second encounter mosaic of the southern hemisphere, the mosaic of the Caloris basin and a unique mosaic of the equatorial regions. The folder labelled "Quad Mosaics" contains scanned images of the mosaics and airbrush maps of the Mercury quadrangles contained in the photographic version of the *Atlas of Mercury* (NASA SP-423) by M.E. Davies, S.E. Dwornik, D.E. Gault, and R.G. Strom.

Each quadrangle imaged by *Mariner 10* contains:

(1) an airbrush map with the latitude/longitude coordinate grid and the names of craters and other features;
(2) a complete image mosaic of the quadrangle; and
(3) enlargements of sections of the mosaic.

A "Coverage Map" shows the quadrangles that were imaged by *Mariner 10*. The folder labelled "Single Images" contains the best *Mariner 10* images of the first, second and third encounters. They are designated by their *Mariner 10* FDS number. The first encounter tape-recorded images (42-59, 27459-27475), and all third encounter images are the raw images as they were received from the spacecraft. They have not been processing because they have little noise. The third encounter images are quarter frames because full frames at the distance and data rate of the spacecraft would have been so noisy that they would have been useless (see Chapter 2). The non-tape recorded images of the first and second encounter are fairly noisy because they were taken and sent back in real time at a relatively low data rate. These images have had their noise suppressed. They are not extensively processed, because users may want to do certain types of processing to suit their own needs. In the folder labelled Raw Images are one raw non-tape recorded first encounter image and a raw second encounter image so you can see the noise level in these images. The quadrangle mosaics have been extensively processed to display all the detail inherent in the images. They do not, however, include the highest resolution images.

SOFTWARE

You can display and do your own enhancement of the images with the included public domain software. All of the images are in TIFF format and can be opened with most image display software. The included software was developed by Wayne Rasband at the U.S. National Institutes of Health. There are two versions, one for Windows and the other for the Apple Macintosh. Documentation accompanies both the Mac and Windows versions. Be sure to read the "Readme" file for a complete explanation of the CD contents and some suggested processing techniques to enhance the images.

If you experience any problems during the installation please compare the technical specifications with the requirements mentioned above. Otherwise please contact our hotline.

Helpdesk electronic media
Springer-Verlag
Tiergartenstr. 17
D-69121 Heidelberg, Germany

Phone (Germany): 0180 565 6665
Phone (international): +49 6221 487 8235
Fax: +49 6221 487 8364
http://www.springer.de/helpdesk-form

1

The twilight planet

1.1 THE PLANET CLOSEST TO THE SUN

A beautiful sight, but today seen by relatively few, Mercury shines brightly in the twilight sky. In our busy modern culture, most people know they can see Mercury with the naked eye only after a purposeful search, having found its location, for the evening or morning from an astronomical source like *Sky and Telescope* or *Astronomy* magazines.

An unusual and wonderful alignment of six planets (including Earth) occurred while this book was being written. Jupiter, Saturn, Mars, Venus, and Mercury graced the western sky for several weeks in late April and early May, 2002. The left image on the frontis piece shows the planetary lineup in stunning beauty. Many backyard observers, thrilled with this viewing opportunity, saw Mercury for the first time.

Early risers may mistake Mercury for a star in the morning twilight. The right image of the frontis piece shows Mercury in the company of stars in the Hyades star cluster as they rise over a Tucson home. The occupant of this house might leave quickly for work, perhaps gazing at the sky and never realizing that Mercury is visiting just above the horizon.

But people of most ancient cultures were more familiar with the twilight wanderers and guardian of the Sun. Because Mercury is the closest planet to the Sun, its hurried orbit takes just 88 Earth days, and switches its appearance from east (morning twilight) to west (evening twilight) in the sky with a slightly variable period of about three months. Mercury also has an *orbital plane* that is tilted with respect to that of the Earth, so it also bobs up and down in the sky relative to the plane of the Earth and the Sun. Thus, sometimes it is lower than the Sun in *declination* and at other times it is higher. These two striking orbital peculiarities further contribute to Mercury's elusiveness.

These motions are especially confusing to people living in mid- and high-latitudes. By far, Mercury is, and has been, most familiar to people living in equatorial and tropical latitudes, especially those with dry climates and clear skies.

1.2   EARLIEST OBSERVATIONS AND RECORDING IN MYTHOLOGY

There is ample linguistic evidence that some of the first people to observe Mercury and commit the planet to the immortality of myth were Germanic people and Scandinavians who navigated far south from their native lands into what is now the Mediterranean and the coasts of Africa. Mercury was connected with the deity Wodan, or Odin among the northern seafarers. In what is now Italy, ancient people called Mercury Boudha, a word with the same origins as Wodan and Odin. The connection lives on in our current use of the English day of week Wednesday derived from "Wodan's day" and present-day Swedes and Danes use of "onsdag", directly derived from the old Norse "Odinsdagr", and also in the French word "mercredi" coming from the Latin "Mercury dies".

The ancient civilizations of the Middle East knew that the time between the reappearance of Mercury in the same configuration in the sky was shorter than for the other planets, and correctly reasoned that it, therefore, moved more rapidly. There was confusion among the early Greeks concerning the *eastern elongation* of Mercury when it is an evening "star" and the *western elongation* when it is a morning "star", which led them to believe these "stars" were two distinct objects. They later recognized that the two "stars" were the same object and the Greek astronomer Eudoxus correctly measured the time interval of 115 days between subsequent appearances in the morning or evening sky in about 400 BC. This period is called the *synodic period*. When Mercury reaches its largest angular distance from the Sun, as seen from Earth, it is said to be at *greatest elongation* (Figure 1.1).

One early, dated observation of Mercury was on 15 November 265 BC, and was recorded in ancient cuneiform writings from Mesopotamia (Greek origin meaning "between the rivers"), one of the earliest centers of urban civilization, in the area between the Tigris and Euphrates rivers. This region today is modern Iraq and eastern Syria. Mercury was given the name "Nebo" – the Mesopotamian god of wisdom and writing. It is said that Nebo was the scribe who kept the book of fate. Other versions of the same god are "Ninib" and "Nabou". Early Christianity adopted Nebo into biblical stories and the name is associated with geographical land forms (Mt. Nebo) and people of the old testament.

Another specific observation was recorded by Chinese astronomers on 9 June 118 AD. Mercury was seen one degree from the star cluster known as Praesepe in the constellation Cancer. During the Middle Ages, when Arab astronomy was flourishing, Mercury was designated "Kokab Outharid" and the Turks called it "Outharid". Again, the association is with a scribe (in some descriptions male, in others female) recording the past, the present, and the future, sitting with a diamond sparkling in her tierra in the twilight and sometimes with four wings to express rapid motion.

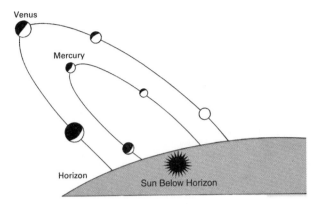

Figure 1.1. As seen from Earth, both Mercury and Venus stay close to the Sun because they orbit the Sun within the Earth's orbit. At their largest angular distance from the Sun as seen from Earth, they are at greatest elongation. In this diagram, the two planets are shown at eastern elongation when they set after the Sun and appear as evening "stars" (from Strom, 1987).

To early Greeks, Mercury was regarded as both male and female like the Sun. Original Grecian dialects used words describing passionate desire, fertility, and rapid movement, with variations conveying the meaning of a tireless god, a traveler, the god of twilight who announced the God of the Day, Zeus. Around the same time, the fascinating element quicksilver with its sparkling sheen, rapidly moving character, and elusive nature resembled characteristics of the planet Mercury, and so the elemental symbol given to the substance was Hg for Hydrargyrum or "water silver" and called Mercury today. Later the name Hermes became a common name for Mercury. The sparkling female goddess association was dropped and a male with winged feet adopted, along with a tireless god, a traveler, protector of trade and procurer of grain. The astronomical symbol of Mercury ☿ can be traced to a medieval Greek manuscript where it takes the form ☿. The horizontal cross is a modern addition. The "horns" at the top of the symbol represent the wings of this speedy planet. It is from the use of the name Hermes for Mercury that the usage of Hermean for characteristics of Mercury became popular during the 19th century and continues to be used by some today. In this book we simply refer to Mercury directly and to Mercury's characteristics.

Depictions of Mercury in its anthropomorphic form illustrate people's continuing fascination with the planet and its legends. Both ancient (Figure 1.2) and modern building façades honor the god and his winged boots.

1.3   EARLY TELESCOPIC OBSERVATIONS

The invention of the telescope ushered in the modern era of astronomy. The telescope was first used by the English scientist Thomas Harriot in August 1609 and later that same year by the Italian astronomer Galileo. Mercury proved to be an elusive object for telescopic observations because it is never more than 28 degrees

Figure 1.2. (Top) Mercury graces the façade of New York City's Rockefeller Center; (Bottom) Mercury and Venus are off on a journey in this fifteenth century BC clay relief from the Sanctuary of Persephone in Lorcri, southern Italy (courtesy of E.C. Krupp, Griffith Observatory).

from the Sun as viewed from the Earth. Observations must be made either at twilight or during the day. The advantage of twilight is good contrast between the illuminated disk of the planet and the relatively dark sky but the disadvantage is that the pathlength of Earth's atmosphere between the planet and the telescope is great; atmospheric refraction is at a maximum and turbulence is often at its worst. Daytime

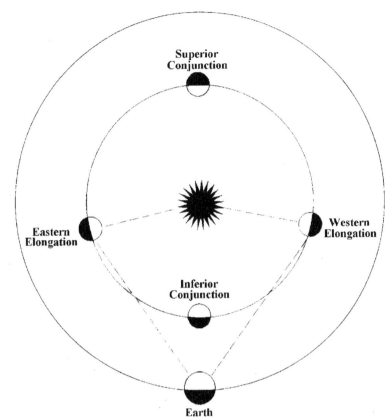

Figure 1.3. This diagram shows elongations and conjunctions of an inferior planet (planets interior to the orbit of Earth). When Mercury or Venus is between the Earth and Sun, it is said to be at inferior conjunction. When they are on the far side of the Sun they are at superior conjunction. Sometimes at inferior conjunction the nodes of the orbital planes align, and Mercury or Venus cross or transit the Sun's disk to appear as dark spots (from Strom, 1987).

observations have the advantage of a much shorter pathlength through the atmosphere, but the contrast between the disk of the planet and the sky is small, and Mercury is often too difficult to find. This was especially true in the past before telescopes benefited from computerized finding and tracking systems.

Compounding these difficulties is the fact that there are only about thirty or forty days a year when Mercury is at an angular distance from the Sun, as viewed from Earth, enabling it to be viewed with a telescope. Additionally, this difficulty is enhanced by the fact that Mercury is an *inferior planet* like Venus, because its orbit is between the Earth and the Sun. As a consequence, the Earth facing disk of Mercury goes through phases much like the Moon. Sometimes when Mercury is at an angular distance from the Sun where it can be viewed from Earth, it has a crescent phase that makes it even more difficult to find (Figure 1.3).

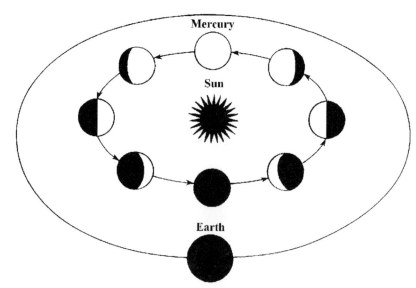

Figure 1.4. In 1639 the Italian astronomer Giovanni Zupus discovered that Mercury shows phases like those of Venus as it orbits the Sun, and like the Moon as it orbits the Earth and the Sun (from Strom, 1987).

In 1639 Giovanni Zupus discovered that Mercury went through phases like the Moon. He used a more powerful telescope than the one used by Galileo who had earlier discovered Venus also went through phases (Figure 1.4). These observations were easily explained by understanding that the two planets orbited between Earth and the Sun. These observations were profoundly significant because they were proof that the *Copernican theory* was correct and the Earth was not the center of the Universe.

Except for the discovery of the phases of Mercury, early observations didn't contribute much to our understanding of the planet. Telescope observations and astronomical drawings of Mercury's surface suffered from the poor viewing conditions and bright sky just as they do today. Early maps of Mercury's surface vaguely resemble Mercury's actual surface in some respects, mostly albedo differences.

Below are some of the more famous results of the early observers. Maps were made complete with Latin names for real and imagined features. Percival Lowell's maps included canal like features (Figure 1.5). We include them here for historical perspective and encourage the reader to appreciate how far technology has advanced in the past 400 years, permitting greatly improved observations that will be illustrated later in this chapter.

By the early 1960s about twenty maps of Mercury had been compiled, some based on extensive visual observations by experienced astronomers such as Eugene Antoniadi, Georges Fournier, and Audouin Dollfus. All appeared to be consistent with an eighty-eight-day rotation period. We know today that Mercury's rotation period is 58.6 days.

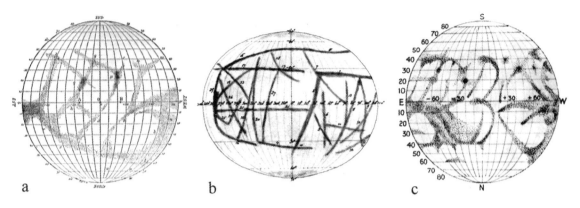

a   b   c

Figure 1.5. (a) Giovanni Schiaparelli's map of Mercury (1889); (b) Percival Lowell's map of Mercury, drawn in (1896); (c) the Rene Jarry-Desloges map of Mercury (1920). Notice south is at the top in these drawings. This is the natural configuration of telescopic images at the eyepiece. Today additional optics in many telescopes are added to invert the image for the convenience of the viewer. Imaging with pixilated CCD detectors and software for "flipping" images is also widely used.

What went wrong with these early estimations of rotation periods? Were the visual observations worthless? It turns out that the visual observations are consistent with both an 88- and 58-day rotation period. A 58.6 day period is exactly two-thirds of the planet's eighty-eight-day year. This means that Mercury rotates three times on its axis every two orbits around the Sun. As a consequence, after three synodic periods we see the same face of the planet at the same phase. Because three synodic periods also corresponds with the time interval between favorable observing conditions – when most visual observations were made – an observer working only at such times will see the same surface features in the same places for six consecutive years. This was precisely the criterion used to justify the 88-day rotation period. This geometry shows shifts over many years but because maps are made over only a few years, it is not surprising that most of an observer's drawings are consistent with one map.

Many people living today were taught incorrectly, in school, that Mercury has a "dark side" that never sees the Sun. This was because early visual astronomers thought that Mercury's rotation period was the same as its orbital period and that the same side of Mercury always faced the Sun much like the same side of the Moon always faces Earth. Older school textbooks carried this misinformation into the early 1970's.

1.4   MODERN TELESCOPIC OBSERVATIONS

In 1965, Gordon Pettengill and Rolf Buchanan Dyce, using the Arecibo radar facility in Puerto Rico, discovered that Mercury's rotational velocity was consistent

Figure 1.6. The small elongated dot in the sky is Mercury, imaged by the *Surveyor 7* lander from the vicinity of Tycho crater on the Moon. It was taken on 23 January 1968. The fuzzy glow on the horizon is the Sun's corona. Mercury's elongated shape is due to its motion during this 30-minute exposure.

with the rotation period of about 59 days. Subsequent radar observations confirmed this result and established that the planet has a rotation period of 58.6 days – not the synchronous rotation indicated by the visual and photographic observations.

In 1968 *Surveyor 7* took an image of the solar corona with a long exposure time. Contained within the picture is Mercury, seen for the first time from another solar system body (Figure 1.6).

While the geometrical and astronomical conditions remain the same, the instrumental conditions for observing Mercury have improved dramatically in the last decades since the advent of CCD (charged couple device) imaging detectors, computerized acquisition and guiding programs for optical telescopes, and rapid data analysis software with the ability to correct for many of the distortions inherent in optical viewing systems. Even relatively small telescopes have made significant progress in imaging Mercury's surface with results far surpassing the finest drawings by experienced observers. Because detectors are very sensitive it is possible to image the planet in a very short time such as 1/1000 of a second. Rapid imaging combined with rapid storage of the image, permits capturing images at moments of minimum atmospheric turbulence and spectra can be obtained with little change in sky conditions.

Amateur astronomers also make valuable contributions to our knowledge of Mercury, especially its surface features. The CCD images of Mercury shown in Figure 1.7, made by members of the Association of Lunar and Planetary Observers (ALPO), Mercury subsection, attests to this claim. Even with a relatively

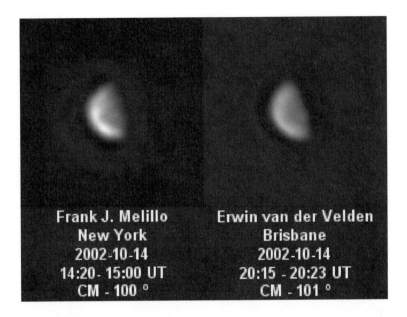

Figure 1.7. These CCD images of Mercury were obtained by members of the Amateur Lunar and Planetary Organization (ALPO) Mercury Section during the October 2002 apparition of Mercury. North, top; West, to right.

large geographic pixel of many hundreds of square kilometers, bright albedo features at locations of fresh craters are clearly seen.

1.5 RESEARCH TELESCOPIC FACILITIES

The major advantage of larger diameter telescopes, besides greater light gathering power, is the larger image scale at the detector. Thus, it is possible to have more pixels across the diameter of Mercury to produce much better spatial resolution of the surface.

1.5.1 Mt. Wilson Observatory

The first real advance in this area was made by a team at Mt. Wilson Observatory in 1998 who used video imaging at the eyepiece of the 1.5 meter telescope to obtain the truly marvelous image of Mercury shown in Figure 1.8.

1.5.2 Swedish Vacuum Solar Telescope

Another effective, but much more difficult technique of obtaining cutting edge images of Mercury is to wait for moments of extremely good viewing with a solar

Figure 1.8. Mercury image obtained from the Mt. Wilson 1.5-m telescope with rapid read-out video equipment and processed with a co-adding algorithm emphasizing sub-pixel alignment. The central longitude of the illuminated sector is ~315°. Mercury is ~7.8 arcsec in diameter (courtesy Jeffry Baumgardner, Michael Mendillo and Jody Wilson, Boston University).

telescope. Johan Warell and colleagues have imaged most of Mercury's surface using the Swedish Vacuum Solar Telescope in the Canary Islands. Many evenings and mornings were spent capturing images during brief periods of atmospheric calm (Figure 1.9).

1.5.3   Future observations

As this book was being written, adaptive optics and active guiding systems are becoming available on many telescopes, both small and large. Very soon far better images of Mercury's surface will become available compared with the ones shown in this chapter. Image stabilization technology combined with the availability of telescopes (1–4 m) for planetary work, provides great opportunities for fully mapping Mercury's surface at about 300 km spatial resolution before *MESSENGER*'s first flyby in 2007.

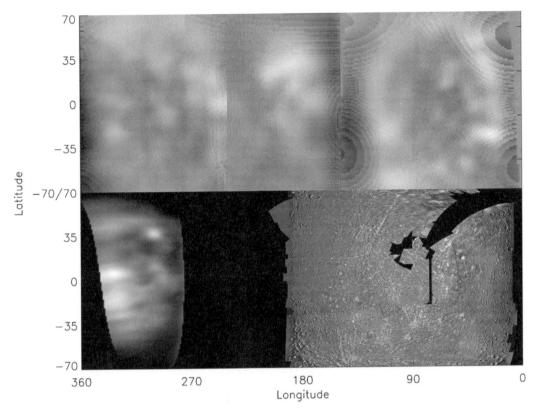

Figure 1.9. State-of-the-art imaging of Mercury, demonstrated by a montage of images taken in 1998–1999 at the Swedish Vacuum Solar Telescope (top panel), and at Mt. Wilson (lower left panel), and in 1974 and 75 by television cameras on board *Mariner 10* spacecraft (lower right panel).

2

The *Mariner 10* mission

2.1 THE ONLY ONE

Mariner 10 is the only spacecraft to have visited Mercury. In the early 1960s a number of planetary missions were being considered, particularly to Mars. However, it was discovered in 1962 that the positions of the Earth, Venus, and Mercury would be configured in such a way in 1970 and 1973 that a spacecraft launched to Venus could be nudged by its gravity field to send it on a new trajectory to Mercury. Thus, with a single spacecraft it would be possible to visit two planets with a minimum expenditure of on-board fuel. The launch had to take place in 1970 or 1973 because the next economical opportunity would not occur until the mid-1980s. Therefore, there was a sense of urgency to devise a mission to Venus and Mercury as soon as possible. In 1968 the Space Science Board of the National Academy of Sciences endorsed a mission to fly by Venus and Mercury in 1973. Late in 1969 Congress approved the mission to begin development in 1970.

2.2 MISSION CONCEPT

It was never intended for this mission to orbit Mercury because it would be traveling so fast past Mercury (~50 km/sec) that it would require a huge amount of fuel to slow down the spacecraft enough to put it into orbit. The size of the retrorocket would have to be equivalent to a medium-sized launch vehicle of that era, and require a launch vehicle comparable in size to the *Saturn V* moon rocket. At that time we did not know about multiple encounters with Venus and Mercury that could slow the spacecraft to low enough speeds to put it in orbit with a relatively small retrorocket. Furthermore, this mission was conceived as a first reconnaissance of Mercury that would be followed in several years by a more sophisticated orbiter. Of course, this is still not the case, and it will have been over 35 years before the next spacecraft explores Mercury.

2.2.1   NASA chose the Jet Propulsion Laboratory

The National Aeronautics and Space Administration (NASA) designated the Jet Propulsion Laboratory (JPL) of the California Institute of Technology to develop and operate the mission. A *Mariner*-type spacecraft was chosen for the mission, and the project was named the Mariner Venus/Mercury Mission. The spacecraft was called *Mariner 10* because it was the 10th *Mariner* to be launched.

Although this spacecraft had to penetrate and operate in a more hostile environment than any previous spacecraft, NASA insisted that it initiate a new breed of low-cost missions. This may sound familiar to the concept of "faster, cheaper, better" that NASA espoused in the 1990s. As with the failures of the two Mars missions in 1998 and 1999, this concept of "cheaper and better" in 1970 almost resulted in the loss of the Mercury mission. One of the most severe restrictions was the use of only one spacecraft to explore two planets. Previous lunar and planetary missions had used two or more spacecraft to gather more data and provide a backup in case one failed. The *Mariner 8* spacecraft to Mars, for instance, experienced a launch failure, leaving only *Mariner 9* to carry out the first orbital mission to that planet. Fears of a similar failure prompted the Mariner Venus/Mercury Mission to request a backup spacecraft when it became evident that one could be prepared within the total project cost of $98 million. Basically, a backup involved only a small increase in the number of spare components. However, this backup spacecraft was to be used only if a failure of the prime spacecraft or launch vehicle occurred during the 16 October to 21 November launch window. If the primary spacecraft failed on or after 21 November, the second spacecraft could not be launched and the mission would be lost. Understandably, the project scientists and engineers were very concerned. Naturally the scientists and engineers wanted to launch both spacecraft so that more data could be obtained, and also in case one spacecraft failed after 21 November. However, to save money they only launched one. As it turned out, this spacecraft was successful, but it almost failed just before it encountered Mercury, as you will see later in this chapter. Today the backup spacecraft is on display at the Smithsonian Air and Space Museum in Washington, D.C.

2.2.2   A necessary gravity-assist

The *gravity-assist* trajectory technique was needed to obtain an economically acceptable mission. This technique allows a spacecraft to change both its direction and speed without using valuable fuel, thereby saving time and leaving more weight for the scientific instruments. Thus, the *Mariner* spacecraft could be launched with an acceptable payload by a relatively cheap *Atlas/Centaur* rocket. If a direct flight to

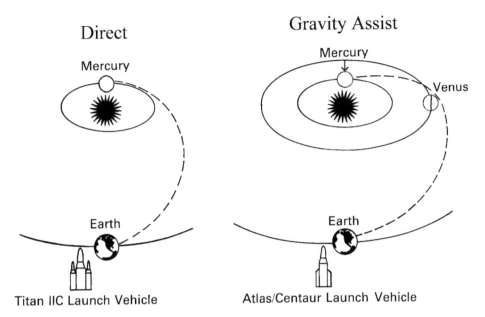

Figure 2.1. The gravity-assist trajectory to Mercury used the gravity and orbital motion of Venus to send the spacecraft into the innermost part of the solar system without the need to expend precious fuel except for minor trajectory corrections. A direct flight without a Venus assist would require a much larger launch vehicle to deliver the same payload. *Mariner 10* was the first planetary mission to use the gravity assist trajectory, which has since been used by other missions such as *Voyager* and *Cassini* for the exploration of the outer solar system (modified from Dunne and Burgess 1978).

Mercury were used, a much larger and more costly *Titan IIIC/Centaur* would be required, and its high speed past Mercury would permit only a short time to acquire data. Not only did this gravity-assist trajectory enable one spacecraft to explore two planets, it also provided a bonus return visit to Mercury (Figure 2.1).

At a conference in February, 1970, on Mercury and the Venus/Mercury Mission at Cal Tech in Pasadena, California, the late celestial mechanician Giuseppe Colombo of the Institute of Applied Mechanics in Padua, Italy, noted that after *Mariner 10* flew by Mercury, its orbital period around the sun would be quite close to twice Mercury's orbital period. He suggested that a second Mercury encounter could therefore be accomplished. A detailed study of the trajectory by JPL confirmed Colombo's suggestion and showed that by carefully selecting the Mercury flyby point, a gravity correction could be made that would return the spacecraft to Mercury about six months later. In fact, it would be possible to achieve multiple encounters with Mercury; the number depending on the fuel available for midcourse corrections and attitude control. *Mariner 10* would eventually achieve three encounters with Mercury before it ran out of fuel.

2.3   THE FLIGHT PLAN

2.3.1   A narrow launch window

The mission plan for *Mariner 10* was the most complex for any planetary mission up to that time. It called for a launch sometime between 16 October and 21 November, 1973. NASA chose 3 November so that the spacecraft would encounter Mercury at a time when it could view the planet about half lit (quadrature). Viewing Mercury at this phase would make it easier to distinguish surface features by their shadows. The trajectory relied on Venus' gravitational field to alter the spacecraft's flight path and speed relative to the Sun. Properly aimed, the spacecraft's speed would be reduced, causing it to fall closer to the Sun and cross Mercury's orbit at the precise time needed to encounter the planet.

New levels of navigational accuracy were required to intercept Venus with high precision. The flyby point at Venus had to be controlled within 400 km, or a Mercury encounter would not occur. At least two spacecraft maneuvers would be needed between Earth and Venus, and probably a further two between Venus and Mercury. *Mariner 10* would be the first planetary mission to use this gravity-assist technique.

The flight plan called for the upper-stage *Centaur* rocket to be turned off for 25 minutes shortly after launch from the Kennedy Space Center. This would place it in a parking orbit that would carry it partway around the Earth. Then a second ignition would thrust the *Mariner* spacecraft in a direction opposite to the Earth's orbital motion, providing the spacecraft with a lower velocity relative to the Sun than the Earth's orbital velocity. This would allow it to be drawn inward by the Sun's gravitational field to achieve an encounter with Venus. After a few months *Mariner 10* would approach Venus from its night side, pass over the sunlit side, and, slowed by Venus' gravitational field, fall inward toward the Sun to rendezvous with Mercury.

2.3.2   Spacecraft design

The *Mariner 10* spacecraft evolved from more than a decade of *Mariner* technology, beginning with the Venus mission in 1962 and culminating with the Mars orbiter in 1971. *Mariner 10* would be the last of the *Mariner* spacecraft to fly. Like the other *Mariners*, it consisted of an octagonal main structure, solar cell panels, a battery for electrical power, nitrogen gas jets for three-axis attitude stabilization and control, star and Sun sensors for celestial reference, S-band radio (12.6 cm wavelength) for command and telemetry, a *high-* and *low-gain antenna*, and a *hydrazine* rocket propulsion system for trajectory corrections (Figure 2.2, see colour plate section).

Mariner 10 flew much closer to the Sun than any previous spacecraft. It was subjected to solar intensities up to 4.5 times greater than at Earth, requiring thermal control to maintain temperatures at a level that would not damage the spacecraft systems. To meet this requirement, a large sunshade, louvers and protective thermal blankets, and the ability to rotate the solar panels were added to the design. Because

Mariner 10 would be so close to the Sun, only two solar panels were needed to generate enough electricity to power the spacecraft. As the spacecraft approached the Sun, the panels could be rotated to change the angle at which light fell on them, and thus maintain a suitable temperature of about 115°C.

Another major design change from past *Mariners* was the addition of a capability to handle up to 118,000 bits per second of imaging data. If this high data rate had not been implemented, all television pictures of Mercury would necessarily have been recorded on the tape recorder and played back at a later time, as occurred on previous planetary missions. Since the tape recorder was capable of holding only 36 images, the amount of high-resolution coverage of Mercury taken during the short interval near closest approach would have been severely limited. The high data rate permitted the spacecraft to transmit high-quality pictures in "real time", or as fast as they were taken, thereby allowing hundreds of images to be obtained during the missions critical encounter phases. This capability provided 5 times as many images and much greater high-resolution coverage than would otherwise have been possible. The high data rate capability turned out to be crucial to the success of the second and third encounters, because the tape recorder failed after the first flyby. Today we fly reliable solid-state storage devices with large capacities rather than tape recorders.

2.3.3   Scientific payload

Since so little was known of Mercury, the scientific instruments to be flown by *Mariner 10* had to be chosen carefully in order to explore the planet as thoroughly as possible. To this end, seven experiments were selected: television imaging: *infrared radiometry*; *ultraviolet spectroscopy*; magnetic fields; *plasma science*; charged particles; and radio science. Radio science did not require a special instrument because it uses the onboard radio system. These experiments would provide data to explore the interior, surface, and near-planet environment of Mercury, and also to obtain data on the atmosphere and space environment of Venus.

2.3.3.1   *The television experiment*

The imaging system consisted of two *vidicon cameras*, each with an eight-position filter wheel. The vidicons were attached to long focal-length *Cassegrain telescopes*, which were mounted on a scan platform for accurate pointing. These telescopes provided narrow-angle, high-resolution images, and were powerful enough to read fine newsprint from a distance of a quarter of a mile. They were absolutely essential for the study of Mercury's surface.

A principal concern of the atmospheric scientists on the Imaging Science Team was the inability of the narrow-angle cameras to image large portions of Venus with one image around the time of encounter. To study atmospheric circulation it is desirable to take pictures of large portions of the atmosphere over a relatively long period, even if the images are at relatively low resolutions. Since this flight would be the first time a spacecraft imaged the atmosphere of Venus, atmospheric scientists desperately wanted such pictures. The budget for this project, however, was extremely tight, and any system to obtain such pictures had to be inexpensive and not interfere with the narrow-angle optics. One evening after an Imaging Science Meeting in Pasadena, California, several team members were discussing this problem over cocktails at a local restaurant. During this discussion, Bruce Hapke suggested an optical design (which Verner Suoumi and Michael Belton sketched on the back of a cocktail napkin) that could provide the equivalent of wide-angle cameras within the budget. The design consisted of auxiliary optics attached to each camera and could be operated by moving a mirror on the filter wheel to a position in the system's optical path. The next morning this design was presented at the team meeting at JPL, and eventually it was incorporated into the camera system.

The primary objective of the imaging experiment was to study the physiography and geology of Mercury's surface, determine accurately its size, shape, and rotation period, evaluate its photometric properties (the manner in which light is reflected from its surface), and search for possible satellites and color differences on its surface. The cameras also took pictures of Venus to determine its cloud structure and atmospheric circulation. Five filters on the two different cameras were used at Mercury. The effective central wavelengths of the filter band passes were: clear (CLR, 487 nm); ultraviolet (UV, 355 nm), blue (BL, 475 nm); minus ultraviolet (MUV, 511 nm); and orange (OR, 575 nm). In addition to wide-angle capability, the filter wheels included a UV polarizer for Venus observations and a calibration lens.

2.3.3.2   Infrared radiometer

The infrared radiometer measured the thermal emission from Mercury and the clouds of Venus using two broad wavelength bands centered at 11 and 45 µm. Brightness temperatures provided information on the thermal properties of Mercury's surface material and were used to infer surface roughness, size of the particles that make up the surface, and whether or not there were rock outcrops and if so, their size. The thermophysical properties of the top few centimeters of the surface were inferred from the rate of cooling on the evening side of the planet. The spatial resolution of the observations was as small as 45 km.

2.3.3.3   Extreme ultraviolet (EUV) experiments

Theoretical predictions indicated that the most likely constituents of an atmosphere on Mercury would be hydrogen, helium, carbon, oxygen, argon, and neon. Even carbon dioxide was a possibility. Consequently, extreme ultraviolet spectrometers were designed to detect these elements. Two instruments sensitive in the extreme ultraviolet (300–1657 Å) were designed and flown on the spacecraft. The airglow spectrometer (polychromator) was placed on the scan platform and the occultation grating spectrometer was mounted on the spacecraft body.

The instrument called the occultation spectrometer was designed to make measurements of the Sun as it passed behind the *limb* of the planet. Any extinctions of sunlight above the limb of Mercury could be attributed to gases in the atmosphere. Channel electron multipliers measured the solar flux at four wavelength positions chosen to cover the first ionization bands of Ne, He, Ar, and Kr.

The second instrument was a spectrometer designed to search for airglow at wavelengths specifically chosen as possible sources: H, He, He^+, Ar, Ne, O, Xe, and C. There were 10 airglow channels and 2 control channels. This instrument was on the scan platform so that measurements could be made across the disk of the planet.

Observations of the bright side of the Moon and Mercury were made with the airglow spectrometer and obtained the first and only measurements of Mercury in the EUV. In addition, Venus, and hydrogen and helium radiation emanating from outside the Solar System were observed with the spectrometer.

2.3.3.4 Magnetic field experiment

The purpose of this investigation was to study any possible magnetic field environment around Mercury and the nature of the *solar wind* interaction with the planet. The magnetic field experiment consisted of two triaxial fluxgate magnetometers located at different positions along a 6-m (20 ft) boom extending from the spacecraft. The spacecraft itself generated a magnetic field, so it was necessary to place the sensors at different distances from the spacecraft to measure this field and then to subtract it from the interplanetary field and any magnetic field associated with Venus or Mercury. There was onboard verification of the assumption that the magnetic field of the spacecraft solar array panels was negligible. The spacecraft itself had a variable magnetic field, although this was small compared to the fields found to be associated with Mercury. More about the particles and fields instruments will be found in Chapter 5.

2.3.3.5 Plasma science experiment

To understand the interaction of the solar wind with Mercury, it was necessary to observe the velocity and directional distribution of positive ions and electrons in the solar wind. Two plasma detectors were therefore located on a motor-driven platform attached to the spacecraft body. This experiment showed whether the solar wind interacted with Mercury in a manner analogous to the Earth rather than to the Moon or Venus, thus indicating an intrinsic magnetic field. Thus, the plasma detectors strongly complemented the measurements of the magnetic field discovered by the magnetometer.

The charged particle experiment was designed to observe high-energy charged particles (atomic nuclei and electrons) over a wide range of atomic numbers and energies. The objectives of this experiment included the determination of the effects of the Sun's extended atmosphere (heliosphere) on cosmic rays entering the Solar System from elsewhere in the galaxy, and the search for charged particles in the vicinity of Mercury.

2.3.3.6   Radio and telemetry system

Finally, the radio waves emitted by the spacecraft telemetry system were mathematically analyzed to determine the gravitational effects of Mercury on the predicted trajectory of the spacecraft. In this way, it was possible to accurately measure Mercury's mass and radius. These data provided a means of accurately determining Mercury's density and, hence, estimates of its internal constitution and structure. Gases in an atmosphere refract and scatter a radio signal, and by measuring these effects it is possible to calculate atmospheric pressures and temperatures. An occultation experiment sought to observe changes in the radio waves as they moved through the atmospheres of Venus and Mercury when *Mariner 10* passed behind the planets as viewed from Earth.

2.4   *MARINER 10* GOES TO MERCURY

Finally, all was ready and the launch day approached. The finalized configuration was tested and mounted on the top of the *Atlas/Centaur 34* launch vehicle on Complex 36's Pad B of Cape Canaveral about 10 days before the 3 November, 1973 launch date.

2.4.1   Launch

After numerous tests of the spacecraft and science instruments under the simulated hostile conditions expected on the Venus/Mercury mission, all was ready for this epic journey of exploration. The spacecraft had to be launched on 3 November, 1973, during a short 1.5 hour period. At 12:45 a.m. Eastern Standard Time, *Mariner 10* was sent aloft (Figure 2.3, see colour plate section).

For the first time on any planetary mission, the science instruments were turned on soon after launch. The purpose was to calibrate them in the well-known environment of the Earth–Moon system. Images of Earth were taken for comparison with Venus, and pictures of the Moon were taken for comparison with Mercury. At this time the first of many problems to plague this historic mission occurred. Heaters designed to hold the television optics at temperatures of 4° to 15°C failed to operate. It was feared that the temperatures would drop to a level low enough to affect the sensitive optics and distort the images. Fortunately, the temperature stabilized at an acceptable level and the cameras maintained their sharp focus. Pictures of Earth showed complex cloud patterns in about the same detail expected at Venus (Figure 2.4). They could provide valuable comparisons with the Venus clouds. The spacecraft's trajectory took it over the Moon's north pole, where pictures provided the basis for subsequently improving the lunar cartographic network and extending it more accurately to the far side – a prelude to a similar application for constructing maps of Mercury (Figure 2.5). Plasma, ultraviolet, and magnetic measurements were also made within the Earth–Moon system.

On 13 November the first midcourse correction was successfully executed, and

Figure 2.4. This picture of Earth was taken by *Mariner 10* 6 November, 1973 from a distance of 1.6 million kilometers. Most of the image shows the eastern Pacific Ocean. This was the first time our planet was imaged from farther away than the Moon.

by 28 November it was known that a second correction would be necessary to achieve the required trajectory past Venus. However, the mission plan had always included two maneuvers before reaching Venus. By this time, the launch window had closed and the backup spacecraft, waiting in case *Mariner 10* failed, could not be launched. *Mariner 10* was now completely alone. If it experienced a catastrophic failure after this time, the exploration of Mercury could not take place again for a decade.

2.4.2 Trouble begins

To the shock of scientists, engineers, and operations personnel, *Mariner 10* began experiencing serious problems just after the launch window closed. When the gyros were turned on to roll the spacecraft through a calibration maneuver, the flight data

Figure 2.5. On route to Venus and Mercury *Mariner 10* took this photomosaic of the Moon on 3 November, 1973 to test the performance of the television cameras. It shows a portion of the front and farside (courtesy Mark Robinson, Northwestern University).

system, which kept track of spacecraft events, automatically reset itself to zero. Although this was not a serious problem, it suggested something might be wrong with the power system. Then on Christmas Day, part of the high-gain antenna's feed system failed and caused a significant dip in the signal power emitted by the antenna. Testing indicated that a joint in the feed system may have cracked due to temperature changes. If this problem persisted, no real-time images could be transmitted, and most of the best pictures planned for the Mercury exploration would be lost. The antenna healed itself, and then failed and healed itself again two more times between 25 December, 1973, and 6 January, 1974.

When the gyros were turned on for another roll calibration maneuver, the flight data system did not reset itself as it had done before. The spacecraft appeared to be behaving neurotically. Then on 8 January, the spacecraft automatically and irreversibly switched from its main power system to its backup system. If the backup failed, the mission would be over. From this point on, extreme care was taken in changing the power status of the spacecraft.

2.4.3   Systems restored

Finally, some good news for a change. On 21 January the second midcourse correction was successfully completed. This maneuver was required to make *Mariner 10* fly through a 400 km diameter area about 16,000 km to the right and in front of Venus, as seen from the approaching spacecraft. Failure to achieve this manouver would mean *Mariner 10* would not continue on to its rendezvous with Mercury. Analysis of the trajectory showed that *Mariner 10* would fly within 27 km of the aim point – a magnificent achievement comparable to hitting a dime with a bullet fired from a distance of about 12 km. At this point, all the science equipment was working well, and even the heaters for the cameras came back on by themselves.

On 28 January another near-calamity struck. During a series of spacecraft roll calibration maneuvers, a gyro-induced instability caused the expulsion of attitude control gas at a dangerously high rate. Without this gas it would not be possible to keep the antennas pointed at Earth, and contact with the spacecraft would be lost forever. Before the problem could be corrected, about 16% of the gas was lost. This would prove costly when time came to do subsequent Mercury encounters.

2.4.4   Venus flyby

Despite these problems, *Mariner 10* made its closest approach to Venus at about 10 a.m. Pacific Standard Time on 5 February, 1974. It took more than 4,000 pictures of Venus' atmospheric structure and circulation patterns between 5 February and 13 February, and acquired a wealth of new information about its atmosphere and environment (Figure 2.6). The ailing spacecraft now headed for its primary target, Mercury, which it would encounter 43 days later. But before it could reach Mercury another midcourse correction was necessary.

Mariner 10's troubles were still not over. On 18 February the spacecraft lost celestial reference on the star Canopus. Apparently its star tracker had locked onto a small particle that had drifted off the spacecraft. By the time *Mariner 10* reacquired Canopus, the gyros had been on for 1 hour and 48 minutes, causing more precious attitude control gas to be lost. As the spacecraft neared the Sun it became hotter and hotter, and more particles drifted from the spacecraft, causing the star tracker to lose Canopus lock frequently. More gas was lost. In response, the operations team on earth devised an ingenious method of conserving attitude control gas, called "solar

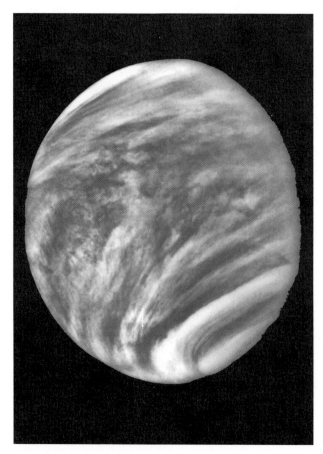

Figure 2.6. This photomosaic of Venus, taken through the *Mariner 10* ultraviolet filter, shows the structure of sulfuric acid clouds in the upper atmosphere.

sailing". By differentially tilting the solar panels to use solar photon pressure on the panels in a controlled fashion, it proved possible to significantly reduce the amount of gas needed for a celestially controlled cruise mode. This and other techniques conserved enough gas to accomplish two subsequent Mercury encounters. Without this technique there would have been only one encounter.

The trajectory past Mercury had been carefully chosen to ensure that the best possible science data could be gathered, and also to allow the spacecraft to return to the planet six months later. This plan required a flyby on the night side at an altitude of about 900 km above the surface. Due to the precise aiming at Venus, only one more midcourse correction was needed to change the flyby from the planet's sunlit side to its night side. Because the planned trajectory correction maneuver would have caused the loss of too much precious gas in gyro oscillations, it was decided to execute a Sun-line maneuver that would not require the gyros. In mid-March *Mariner 10*'s position and orientation were such that its rocket engine could be fired toward the Sun without having to roll or pitch the spacecraft. By applying the proper amount and direction of thrust at the right time, the spacecraft was pushed slightly away from the Sun to fly by the night side of Mercury (Figure 2.7, see colour plate section). On 16 March the maneuver was successfully completed, but the flyby was 200 km closer to Mercury than planned. Since this still satisfied the requirements for science and a Mercury return, no additional maneuvers were planned.

The day after the maneuver, the non-imaging science experiments were turned on in preparation for the encounter. Now some truly good news cheered project personnel, particularly the Television Science Team. The high-gain antenna had miraculously recovered and was able to transmit its signal at full strength. Apparently the crack in the antenna feed closed when the temperature rose as the spacecraft approached the Sun. Now most pictures could be taken in real time. The high-resolution coverage originally planned could now be accomplished.

2.5   THE FIRST MERCURY ENCOUNTER – MERCURY I

2.5.1   The first images of the unknown planet

The first pictures of Mercury were taken on 24 March from a distance of 5.3 million kilometers. They were initially about the same quality as pictures taken from Earth, but as *Mariner 10* neared Mercury the images showed a heavily cratered surface superficially resembling the Moon's. The picture-taking sequence called for a series of photographic mosaics to be taken of the half-lit hemisphere as the spacecraft approached the planet. Near closest approach a series of individual high-resolution pictures was taken near the terminator (the line that separates day and night) (Figure 2.8).

Mariner 10 reached closest approach to Mercury at 1:47 p.m. Pacific Daylight Time on 29 March, 1974. For a short time around closest approach, the Earth was occulted by Mercury, cutting off the spacecraft radio signal. All science data, including the highest-resolution television images, were placed on the tape recorder for later playback. As the spacecraft receded from Mercury, a series of image mosaics similar to those obtained on approach were taken of the other side of the half-lit planet.

The pictures revealed a heavily cratered surface similar to the lunar highlands. An enormous impact basin about 1,300 km in diameter was revealed half illuminated at the terminator. There were also large expanses of smooth, lightly cratered plains that resembled the Moon's maria. All these surface characteristics were similar to the now-familiar features seen on the Moon. There were, however, several aspects of Mercury's surface that differed significantly from the Moon's. Long, sinuous cliffs or scarps traversed the surface for hundreds of kilometers and appeared to be almost everywhere. The heavily cratered regions contained large areas of moderately cratered plains interspersed among clusters of craters. A large region of hilly and lineated ground – nicknamed the "weird terrain" – was discovered on the incoming side viewed by *Mariner 10*. These features would enable scientists to reconstruct a

Figure 2.8. As *Mariner 10* approached Mercury, more details could be observed. The image at far left, taken 24 March, 1973, from a distance of 5.4 million kilometers, shows about the same amount of detail as the best Earth-based images obtained at that time. The right-hand image taken 5 days later from a distance of about 1 million kilometers, shows Mercury has a cratered surface.

geologic history of Mercury that was similar to the Moon's in some respects, but significantly different in others (Figure 2.9).

2.5.2   The first real surprise

Because Mercury was not thought to possess a significant magnetic field, it was generally believed that its interaction with the solar wind would be quite similar to that of the Moon, where the wind impinges directly on the surface and the satellite causes a cavity in the wind behind it. At the Earth, Jupiter, Saturn, Uranus, and Neptune, the solar wind is held away from the surface by their magnetic fields. *Mariner 10*'s trajectory would carry it through the anticipated plasma cavity behind Mercury. To the astonishment of scientists monitoring the plasma data telemetered from the spacecraft, 19 minutes before closest approach the plasma flux suddenly increased and peaked in a manner indicating that *Mariner 10* had crossed a *bow shock wave*. At about the same time, the charged particle experiment detected a violent increase in energetic charged particles, confirming that the space-craft had crossed a shock wave. Several other unusual peaks in the intensity of

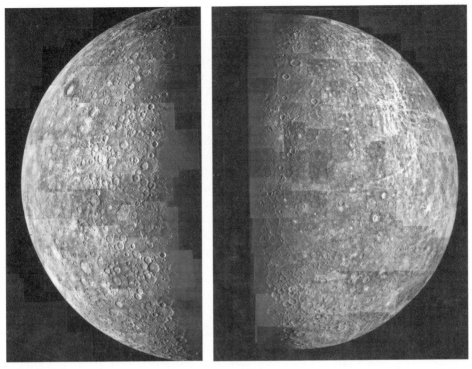

Figure 2.9. These hand-made photomosaics of Mercury were made from medium-resolution images taken on the first encounter. The one on the left shows the incoming side as viewed by *Mariner 10* and the other shows the outgoing side.

charged particles subsequently occurred. They were similar to phenomena observed in the Earth's geomagnetic tail, where oppositely directed field lines meet.

Scientists were both elated and perplexed over this completely unexpected discovery. What could be the source of this magnetic field? Was it internally generated or was it due to electric currents induced in the surface by the solar wind? Another Mercury encounter would be required to answer these questions (see Chapter 5).

2.5.3 A very thin atmosphere, no molecular species found

The ultraviolet experiments put upper limits on molecular and atomic gaseous species at Mercury and determined that the planet possessed a very thin atmosphere of H, He, and perhaps O. More atomic species have since been discovered and will be discussed in Chapter 6. Also, the light emitted in the inner solar system by H atoms (Lyman-α at 1216 Å), much of it coming from H in the solar wind, was measured and mapped.

2.5.4   Hot, hotter, hottest

The infrared radiometer measured a low temperature of −183°C (−297°F) on the night side just before dawn, and a high temperature of 186°C (368°F) in the late afternoon. When Mercury makes its closest approach to the Sun, however, the temperature range can reach 610°C (1,130°F). This enormous temperature difference between night and day is greater than on any other planet or satellite in the Solar System. The temperature gradient between Mercury's day and night side showed that its surface consists of a light, porous insulating layer of dust similar to that on the Moon. Slight temperature variations on the night side indicated the presence of small rock outcrops, probably due to boulder fields around impact craters.

Radio tracking of *Mariner 10* provided an accurate radius of the planet and showed that it is much closer to a sphere than either Earth or Mars. Its mass was measured to an accuracy 100 times greater than previous Earth-based determinations. These values could now provide an accurate density of Mercury, which, in turn, would yield information on its bulk composition and internal structure.

Mariner 10's first encounter with Mercury was an outstanding success that exceeded all expectations. The achievement was particularly noteworthy because of the numerous spacecraft problems that had to be overcome and that had threatened to end the mission before it accomplished its objective. *Mariner 10* provided us with our first close-up glimpse of Mercury and returned thousands of pictures and tens of thousands of non-imaging measurements of its surface and environment. A few hours of spacecraft observations had obtained more data about this poorly known planet than centuries of Earth-based observations. They revealed a planet with a combination of Earth-like and Moon-like characteristics that provided important new information on the evolution of the Solar System.

But this was no time to bask in the glories of the success. A second and possibly a third encounter with Mercury were possible, to add to and complement the knowledge acquired during the first. It was now up to the engineers and operations personnel at JPL to guide the ailing spacecraft to further Mercury flybys.

2.6   THE SECOND ENCOUNTER − MERCURY II

2.6.1   Troubles are overcome

Several additional trajectory corrections were required to return the spacecraft to Mercury. But *Mariner 10* continued to experience serious problems that threatened to terminate the extended mission. Only two days after the first encounter, while the cameras were still taking far encounter pictures, the temperature in the power electronics compartment rapidly rose and was accompanied by an additional 90-watt drain on the power system. The spacecraft remained on its backup power system, and if it failed the mission would be over. The operations team therefore turned off the cameras and other power-consuming instruments, and implemented additional techniques to accommodate the stress on the power system. This response seemed to stabilize the system, but other problems followed.

Without command, the tape recorder turned on and off several times and then failed completely. Without it, all science data would have to be sent back in real time during subsequent Mercury encounters. The high-gain antenna, which had experienced previous problems, would need to transmit at full strength or most of the subsequent pictures would be lost. To make things worse, the flight data subsystem experienced a failure that terminated many of the engineering data channels. This failure greatly increased the difficulty of nursing the ailing spacecraft around the Sun before encountering Mercury for the second time. Finally, the amount of attitude control gas was now quite low because of the oscillation problems, and, therefore, its use would have to be drastically reduced below the normal cruise rate if two more encounters were to be achieved. At this point it looked as if subsequent encounters with Mercury would have only a slim chance to succeed.

Despite these seemingly insurmountable problems, the exhausted personnel at JPL Mission Operations managed to keep *Mariner 10* operating and to preserve enough attitude control gas to accomplish two more Mercury encounters. Because Mariner 10's orbital period around the Sun was almost exactly twice Mercury's period, and since Mercury's rotation period is in $\frac{2}{3}$ resonance with its orbital period, the spacecraft would view exactly the same side of the planet on the second and third encounters as it did on the first.

2.6.2   Conflicts over experiments and spacecraft control

A serious conflict emerged between the magnetic fields and charged particles experiments and the imaging experiment over the trajectory past Mercury for the second encounter. The fields and particles experiments required a night-side trajectory close to the planet, to obtain information needed to determine whether the magnetic field was internally or externally produced. The imaging experiment needed data on the south polar region, to link the two sides of Mercury seen on the first encounter and to determine the polar distribution of the *lobate scarps* in order to decide whether they were produced primarily by *planetary despinning* or by cooling of the interior. This objective required a day-side trajectory at a relatively large distance from the planet. The imaging trajectory would provide little useful information on the magnetic field and interacting particles, while the fields and particles trajectory would yield no new information on Mercury's surface. This was an extremely difficult conflict to resolve. After an agonizing evaluation of both cases by the Science Steering Group, it was decided that the second encounter would take an imaging trajectory, and that the third would be planned primarily for fields and particles. This decision was largely based on the improbability of achieving an adequate imaging trajectory on the third encounter if the second encounter followed a fields and particles trajectory.

2.6.3   Decisions were made

Mariner 10's second encounter with Mercury (known as Mercury II) was chosen to take place on the sunlit side at an altitude of 50,000 km over the southern hemisphere, 40° below the equatorial plane. This trajectory permitted pictures to be taken of a previously unimaged region and provided a photographic tie over the southern hemisphere between the sides of the planet viewed during the first encounter. A rather large miss distance was required to completely cover the southern hemisphere with narrow-angle pictures at resolutions of 1–3 km. Furthermore, the trajectory would enable a third and final encounter with Mercury. The second encounter was designed primarily as an imaging flyby to photograph new areas and to provide important geologic and cartographic links between the two sides seen previously, thereby facilitating further geologic interpretation. An added bonus would be the acquisition of stereoscopic coverage by combining the first and second encounter pictures of the same regions taken at different viewing angles.

2.6.4   Arrival at Mercury

After two more successful midcourse corrections, *Mariner 10* encountered Mercury for the second time at 1:59 p.m. Pacific Daylight Time on 21 September, 1974. The scientists and engineers were jubilant.

About 360 images were returned during the 3-day encounter sequence. They showed details of the south polar regions never seen before and extended the coverage of Mercury from about 50 to 75% of the illuminated hemisphere (see Figure 2.12). The south polar regions revealed a cratered surface similar to that seen on the first encounter (Figure 2.11). An important observation was the presence of numerous sinuous scarps, first seen on the previous encounter, which indicated that these structures have an extremely widespread distribution, possibly on a global scale. This fact has important implications for interpreting the planet's tectonic history and internal dynamics. This similarity of the terrain in the south polar regions with the terrain on other parts of Mercury increased the suspicion that the unseen side was similar to the explored side.

The ultraviolet experiment was able to obtain excellent data. It set accurate upper limits on the density of several molecular species that had been predicted, and detected atomic species of H and He. Two modes of observations were made: (1) an *occultation experiment* searched for molecular absorptions in the solar spectrum; and (2) UV emission lines were searched for in especially chosen wavelength intervals. More about these issues will be found in Chapter 6.

2.6.5   Some historical precedents set

The second encounter was historic because it was the first time any spacecraft had returned to its target planet. Furthermore, engineers at the Goldstone Tracking Station in the Mojave Desert of California were successful in developing a new technique to obtain relatively low noise pictures, required because *Mariner 10* was

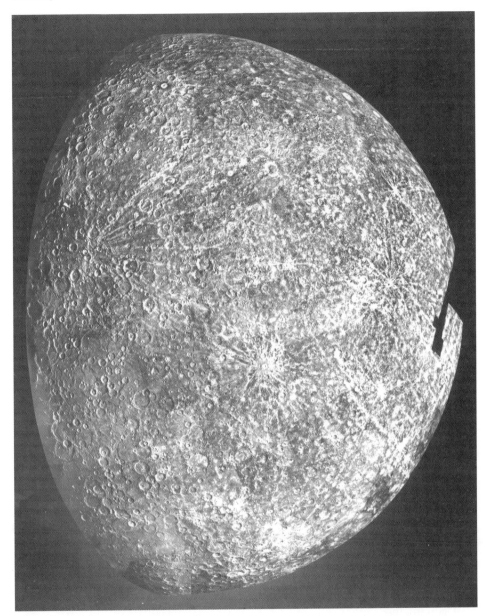

Figure 2.11. This hand-made photomosaic of Mercury's southern hemisphere was made from images taken during the second flyby.

at a much greater distance from the Earth than during the first encounter, and, therefore, its signal was much weaker. They connected three large antennas by microwave links and operated them as a single large antenna. Without this technique, the quality of the full-frame images would have been so degraded that

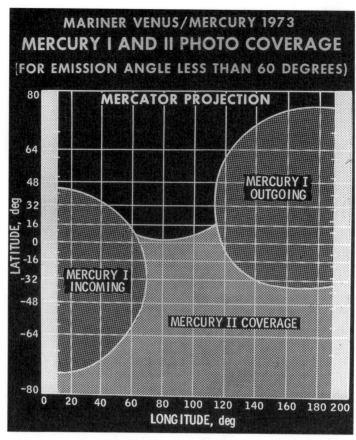

Figure 2.12. The additional coverage of Mercury obtained on the second encounter joined the two sides seen on the first encounter.

they would have been of little scientific value, or it would have been necessary to transmit only $\frac{1}{4}$ of each image at a lower telemetry rate. In either case, most of the objectives of the second encounter would have been lost. Thanks to the ingenuity of the Goldstone engineers, a full complement of excellent images was obtained in real-time. Although these, like the real-time first encounter images, have some noise, it could be removed by image processing techniques.

Up until this time all interplanetary flights had relied solely on Earth-based radio measurements for navigation. In the second *Mariner 10* encounter, a new navigational technique was tested that used the stars as celestial reference points. This navigation method was not unlike that used by ancient mariners to guide them over the vast seas in their explorations of Earth. More than 100 pictures of star fields were taken to obtain angular measurements between Mercury, the spacecraft, and the stars. The experiment was successful, and it demonstrated that long missions to the outer planets could use this method to navigate a spacecraft through the intricate orbits of the outer planets' satellites. *Voyagers I* and *II* would later use this technique on their historic explorations of the outer Solar System.

2.7 THE THIRD ENCOUNTER – MERCURY III

As *Mariner 10* began its second orbit around the Sun in preparation for the third encounter (Mercury III), the spacecraft was returned to cruise mode. The high-gain antenna and solar panels were again used to gather light pressure from solar photons in order to conserve attitude control gas, which was now dangerously low.

2.7.1 Loss of the star tracker

On 6 October the Canopus star tracker again lost its lock on the star when a bright particle passed through its field of view. The spacecraft went into an uncontrolled roll that could not be corrected before the attitude control gas was thought to be depleted below that required to achieve the third encounter.

The situation was now desperate. What could be done to save the remaining attitude control gas for the crucial third encounter? The operations team decided to abandon roll axis stabilizations and permit the spacecraft to slowly roll, the rate controlled by differentially tilting the solar panels. The rates had to be very accurately controlled to prevent excessive use of the pitch and yaw jets, and this was made extremely difficult by the earlier loss of many of the engineering channels. The continuous rolling of the spacecraft also made navigation much more complicated. Despite these problems, however, three more trajectory correction maneuvers were successfully completed that placed the spacecraft on a path that would take it closer to Mercury (327 km) than any previous planetary flyby.

The measures taken to preserve the attitude control gas seemed to be working, and it looked as if there would be enough to achieve the third encounter. Then, a few days before encounter another problem occurred that nearly ended the mission there and then. While trying to reacquire the reference star Canopus, the spacecraft rolled into a position where the signal strength from the low-gain antenna plummeted to a level that essentially broke communications with Earth. If communications with the spacecraft were not reestablished soon, *Mariner 10* would fly by Mercury in utter silence. Only the large 64-m antennas of the Deep Space Tracking Network were capable of emitting a signal strong enough to command the spacecraft to reacquire Canopus. But these antennas were currently being used to communicate with the *Pioneer* and *Helios* spacecraft.

Time was running out, so a spacecraft emergency was called. To save *Mariner 10*, the big antenna at Madrid, Spain, was directed to send a command to the spacecraft that, it was hoped, would result in the reacquisition of Canopus and the positioning of the spacecraft to resume normal communications. Even though this was the period of maximum scientific interest for the *Helios* Mission, the Madrid station sent the command, and shortly thereafter *Mariner 10* achieved its correct orientation for the third flyby – just 36 hours before its closest approach to Mercury.

2.7.2   Magnetic field measurements

The primary goal of the third encounter was to obtain measurements of the magnetic field that would determine whether it was internally generated or produced externally by electric currents induced by the solar wind. If the field was internally produced representing a scaled-down version of Earth's magnetic field, then the time expected for events to be observed by *Mariner 10* at encounter could be accurately predicted. The actual time that *Mariner 10* passed through the bow shock, the magnetopause, and the maximum field strength proved to be almost exactly those predicted. Thus, Mercury's magnetic field was thought to be internally generated and similar in form to the Earth's field.

2.7.3   Final imaging sequence, diminished but important

Although the non-imaging science instruments were returning important new results, the imaging science experiment was experiencing serious difficulties. Only the Canberra station of the Deep Space Network was within receiving view of the spacecraft during the third encounter. Earlier, an experimental ultra-low-noise feed was installed so that high-quality, real-time images could be received at the high data rate. Near encounter, however, the feed developed a leak in its cooling system, and the imaging data had to be returned at a much lower data rate. As a result, only $\frac{1}{4}$ of each picture could be returned in real time. Of course, *Mariner 10*'s tape recorder had failed much earlier, and hence there was no way to store the pictures for later playback to recover the full-frame images. The imaging sequence had been planned to return high-resolution images of geologically interesting areas seen during the first encounter. Although much of the image data was lost, even the strips of high-resolution pictures proved to be important (Figure 2.13).

2.8   MISSION'S END – A JOB WELL DONE, BUT INCOMPLETE

On 24 March, 1975, just one week after the third encounter, *Mariner 10* ran out of attitude control gas. It began tumbling uncontrollably, and communications were lost forever. The mission was over. On its epic journey of discovery, *Mariner 10* had traveled more than 1.5 billion kilometers since it had been launched 506 days earlier. It used for the first time the gravity assist technique to send a spacecraft on a new trajectory, and pioneered many other new engineering techniques. It had transmitted important new information on the atmosphere and space environment of Venus, and provided detailed information about a planet that had baffled scientists for centuries. None of this would have been possible without the Herculean efforts and ingenious devices employed by the project personnel to keep the ailing spacecraft alive. Today *Mariner 10* continues its endless journey around the Sun, returning every 6 months to the vicinity of its prime target, Mercury.

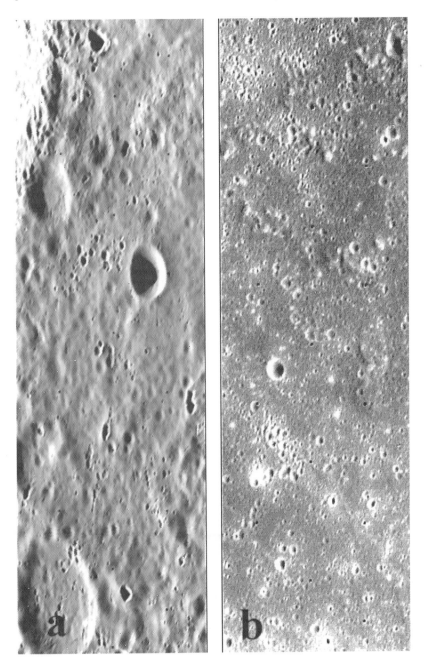

Figure 2.13. The images taken on the third encounter were limited to quarter frames because of ground-based antenna problems. These pictures included high-resolution images of a cluster of sharp secondary craters on the incoming side (a) and a portion of the smooth plains on the outgoing side (b).

3

Mercury's motions

3.1 A DAY ON MERCURY

What would your day be like if you lived on Mercury? It would be totally unlike a day on any other planet, particularly the Earth. As a consequence of the 3:2 resonance, observers on Mercury would witness odd motions of the Sun. During a single *perihelion* passage, observers at the 90 and 270° longitudes would see two sunrises and two sunsets. At perihelion Mercury's orbital speed is so fast compared to its *rotational speed* that an observer at the 90° longitude would witness the Sun rise, hover in the sky, set and then rise once again. An observer at the 270° longitude would see the Sun set, rise again, and then set once more.

At the next perihelion the observer at 90° would see the double sunset and the one at 270° would see the double sunrise. At the zero and 180° meridians an observer would see the Sun rise in the east and climb slowly for $1\frac{1}{2}$ earth months until it reached noon. Just before noon the Sun would appear to perform a loop in the sky – slowing down, stopping, backing up, stopping again, and then continuing westward until it set $1\frac{1}{2}$ Earth months later.

When the planet is at *aphelion*, you would awaken shortly before dawn to find that the surface temperature was at its minimum of $-183°$C ($-300°$F). The Sun would rise in the east and slowly move skyward. At this time the Sun's apparent diameter would be more than twice as large as it is when seen from Earth. But the sky would be black, because Mercury has essentially no atmosphere. After 22 Earth days it would be midmorning, and the surface temperature would have risen to a comfortable 27°C (80°F). Twenty-two Earth days later it would be noon, the Sun at its zenith performing its curious loop in the sky, signaling that you had reached perihelion and completed half a mercurian year. The Sun's apparent diameter would have grown to more than three times its apparent diameter as seen from Earth, and the surface temperature would have climbed to 407°C (765°F). Six Earth days later, in the early afternoon, the surface temperature would reach its

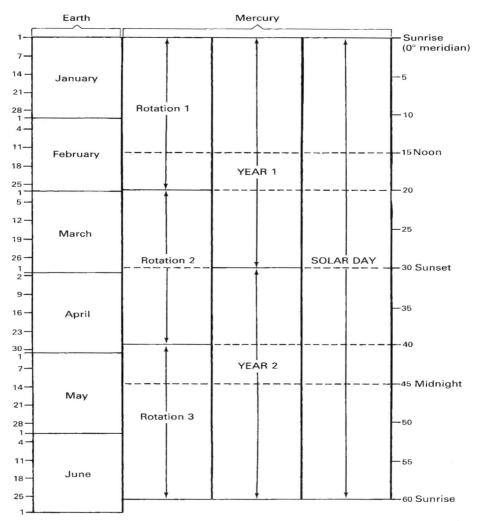

Figure 3.1. If you lived on Mercury you might use a calendar such as this. On this calendar the Solar day would be the longest unit, lasting almost six Earth months (here arbitrarily divided into 60 units). The day would be divided into two Mercurian years, each of 30 unit duration. Each Mercurian year would be divided into 1.5 rotation periods, each rotation period of 20 units duration. This Mercurian calendar is compared to the first six months of an Earth year (from Strom, 1987).

maximum of 427°C (800°F). The Sun would continue its westward journey and finally set 38 days later, when the surface temperature would have fallen to −23°C (−9°F). The sunset would signal that you were again at aphelion and a Mercurian year older (Figure 3.1).

If you are a night owl, you could stay up until midnight, 44 earth days later, to observe the constellation Taurus directly overhead, indicating you were again at perihelion. As you retired for the night you would notice that the surface temperature has continued to fall below $-70°C$ ($-274°F$). When you woke again at sunrise, 44 earth days later, you would again be at aphelion, another Mercurian year older. During your Mercurian day you would have aged two Mercurian years (about $\frac{1}{2}$ earth year), completed three rotations around your planet's axis, and experienced a surface temperature range of $610°C$ ($1,130°F$).

Having described this fantastic day that sounds like it must be something out of science fiction, let us now look at the orbital details that make it real.

3.1.1   A highly eccentric orbit

Mercury has the most elliptical orbit of any planet. (Although Pluto has a more inclined and elliptical orbit it is a Kuiper Belt object; not a planet.) The degree of ellipticity of an orbit is called the *eccentricity*. It is defined as half the distance between the foci divided by the semimajor axis. A perfect circle has an eccentricity of zero and a parabolic trajectory has an eccentricity of 1. Earth's eccentricity is 0.0167 which is close to circular, but Mercury's is 0.205. Therefore, the difference between Mercury's closest point to the Sun (perihelion) and its farthest point (aphelion) is very large. The average distance is 0.387 *Astronomic Unit* (1 AU = 57.9 million km), but at perihelion it is at a distance of 0.31 AU from the Sun, and at aphelion its distance is 0.47 AU – a difference of almost 24 million km (Figure 3.2).

3.1.2   Fleet-footed messenger of the gods

Planetary orbital motions were first described by the legendary Johannes Kepler who formulated the first two laws in 1609 and the third law in 1619 using observations by Copernicus and Tycho Brahe. Kepler's First Law simply states that planets travel along elliptical orbits with the Sun at one focus of the ellipse. Kepler's Second Law states that the speed of a planet is greatest at perihelion and least at aphelion. Kepler's Third Law states that a planet's sidereal period in earth years is proportional to the square root of its semimajor axis (a) cubed ($\sqrt{a^3}$).

The speed of a planet increases as it approaches perihelion and decreases as it gets farther away. Furthermore, since the Sun's gravitational pull is greater the closer a planet is to the Sun, Mercury's orbital speed is greater than that of any other planet. Its average speed is 48 km/s. However, at its closest approach to the Sun it travels at 56.6 km/s. At its most distant point it slows to 38.7 km/s. If an airplane could travel that fast, it would take less than 12 min to circle the Earth. Obviously, Mercury's great speed means that it will complete one orbit around the Sun in a short period of time. This period is called the *sidereal period* (or year) and is described by Kepler's Third Law. Since Mercury's semimajor axis is 0.387 AU, then its year is only 0.24 Earth years, or 87.97 days.

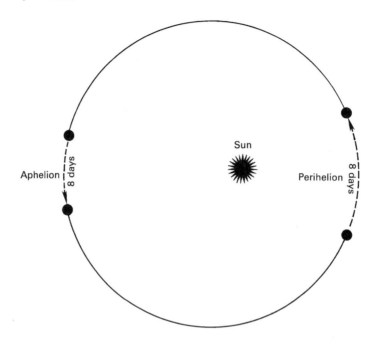

Figure 3.2. Mercury's orbit around the Sun is more elliptical than any other planet (Pluto is a Kuiper belt object). Its eccentricity, however, is still small enough that the orbit appears almost circular. The distance Mercury moves in eight days is shown near perihelion and aphelion. Mercury travels much faster near perihelion (56 km s^{-1}) than near aphelion (39 km s^{-1}) (from Strom, 1987).

3.1.3   Inclined to be noticed

The *inclination* of a planet's orbit is its tilt with respect to the ecliptic plane – the plane of the Earth's orbit around the Sun. Mercury's inclination is 7°, greater than any other planet. We are not sure why Mercury's eccentricity and inclination are so much larger than those of the other planets (Figure 3.3). One suggestion is that its orbit has been stretched out and the inclination increased by perturbations caused by Venus' gravitational field. Another possibility is that large protoplanets were perturbed into the innermost Solar System by close approaches to other protoplanets during the final phases of accretion of the planets. One of these bodies may have collided with Mercury and knocked it into its present eccentric and inclined orbit. This is discussed in detail in Chapter 12.

3.1.4   A slow rotator

Mercury's rotation period is slower than any other planet except Venus. It rotates once on its axis in 58.6 Earth days. This slow rotation and its orbital period account for the very long solar day (sunrise to sunrise).

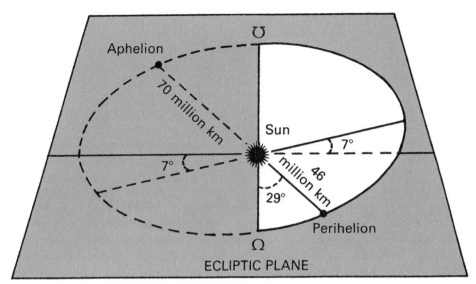

Figure 3.3. Perspective drawing shows Mercury's orbit with respect to the ecliptic plane (the plane defined by the Earth's orbit around the Sun). The ascending (℧) and descending (Ω) nodes represent the points where Mercury's orbit intersects the ecliptic plane. Mercury's perihelion and aphelion points are 29° from the nodes, and its orbit is inclined 7° from the ecliptic plane (from Strom, 1987).

3.1.5 Mercury's strange 3:2 spin:orbit resonance

Since Mercury rotates once every 58.6 Earth days and orbits the Sun in 87.9 Earth days, it rotates exactly three times as it circles the Sun twice. This type of relationship between the rotational and orbital period of a planet or satellite is known as spin–orbit coupling. In the case of Mercury, the rotation/orbit relationship is in a 3:2 resonance or commensurability: it makes three rotations for every two orbits. Our own Moon also has a spin–orbit coupling, because it rotates once on its axis during one orbit around the Earth. The Moon, therefore, has a 1:1 resonance. Most of the outer planet satellites also have this type of resonance.

The 3:2 spin–orbit coupling causes a peculiar diurnal (daily) cycle for Mercury. These combined rotational and orbital motions result in a Mercurian solar day (sunrise to sunrise) lasting two Mercurian years, or 176 Earth days, 88 Earth days of daylight and 88 Earth days of dark. This results from Mercury's rapid speed around the Sun which reduces the effect of its axial rotation. During the time between sunrise and noon, Mercury has completed $\frac{3}{4}$ of its axial rotation, but during the same time it has traveled halfway around the Sun. If the planet had remained where it was at sunrise and completed $\frac{3}{4}$ of its rotation, it would have been midnight instead of noon (Figure 3.4).

Another effect of Mercury's 3:2 resonance is that the same hemisphere always faces the Sun at alternate perihelion passages. This pattern occurs because the

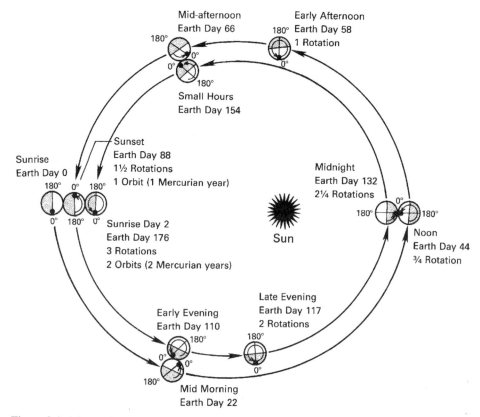

Figure 3.4. Mercury's 3:2 spin–orbit coupling is shown in this diagram, starting at sunrise on the zero degree meridian. After half an orbit, Mercury has rotated $\frac{3}{4}$ of a turn and it is noon at perihelion. After one complete orbit, Mercury has rotated 1.5 times, and it is sunset. At the next perihelion passage, Mercury has rotated 4.25 times, and it is midnight. After two orbits, Mercury has rotated three times, and it is again sunrise. Thus, after two orbits, Mercury has rotated on its axis three times and has experienced one solar day (see Figure 3.1) (from Strom, 1987).

hemisphere facing the Sun at one perihelion will rotate $1\frac{1}{2}$ times by the next perihelion, placing it directly away from the Sun. But at the following (second) perihelion it will have rotated another $1\frac{1}{2}$ times, placing it directly facing the Sun again. The prime meridian (zero degrees longitude) was chosen to pass through the subsolar point at the first perihelion passage that occurred in 1950. Thus, the hemisphere centered on the zero degree longitude will face the Sun at one perihelion and the opposite hemisphere centered on the 180° longitude will face the Sun at the next perihelion. Conversely, at aphelion the hemisphere centered on the 90° meridian will face the Sun, and the opposite hemisphere centered on the 270° meridian will face the Sun at the following aphelion. Thus, the hemispheres centered on the zero and 180° meridians face the Sun at perihelions and the hemispheres centered on the 90 and

270° meridians face the Sun at aphelions. The longitudes with subsolar points at zero and 180° longitude are therefore hot longitudes. The longitudes with subsolar points at 90 and 270° longitude are therefore cool longitudes. Sometimes the 0 and 180° sub-solar points are called the "hot poles", and the 90 and 270° sub-solar points are called the "warm poles."

Mercury's peculiar resonance was apparently acquired over time as a result of the natural consequence of the dissipative processes of tidal friction and the relative motion between a solid mantle and a liquid core. Mercury probably did not always rotate at its present rate. Just after formation it probably rotated much faster and was subsequently slowed by solar tides combined with motions in the liquid core relative to the mantle. Dynamical studies suggest that Mercury may have had an initial rotation period as short as eight hours. At this rapid rotation rate, centrifugal forces would produce an asymmetrical shape with flattened poles and a bulging equator. The Sun's strong gravitational pull would act on the equatorial bulge to produce tidal friction and slow the planet's rotation period. Eventually it would slow to its present rate and become captured into the 3:2 resonance – a dynamically stable configuration. Some of the energy required to reduce the rate of spin of the planet would be converted to heat, thereby increasing Mercury's internal temperature by about 100°C. As Mercury slowed from a rapid rotation rate and centrifugal forces decreased, it assumed a spherical shape.

3.1.6   No seasonal variations

The tilt of a planet's rotation axis to its orbital plane is called its *obliquity*. Seasons are caused by the tilt of a planet's axis of rotation in relation to its orbital plane. The Earth's obliquity is 23.5°. Near aphelion the northern hemisphere is tilted toward the Sun and the southern hemisphere is tilted away. At this time, the Earth's northern hemisphere receives more solar radiation per unit area and summer occurs in the north. At the same time the southern hemisphere is receiving less sunlight and is experiencing winter. Six months later the Earth has traveled halfway around the Sun and the situation is reversed: winter in the northern hemisphere and summer in the southern hemisphere.

Unlike the Earth and Mars, Mercury's axis of rotation is perpendicular to its orbital plane – it has an obliquity of 0°. As a result, Mercury has no seasons. The fact that Mercury's obliquity is 0° means that there are permanently shadowed regions in craters at high latitudes where very cold temperatures are maintained despite the close proximity to the Sun.

3.1.7   When its hot, its hot – when its not, its not

Temperatures on Mercury vary enormously. The planet is only about 46 million km from the Sun at perihelion. At the equator, near noon, the surface temperature reaches 427°C (800°F). This temperature is high enough to melt zinc. At night just before sunrise, however, the temperature plunges to a frigid −183°C, or about −300°F. This represents a temperature difference between day and night of

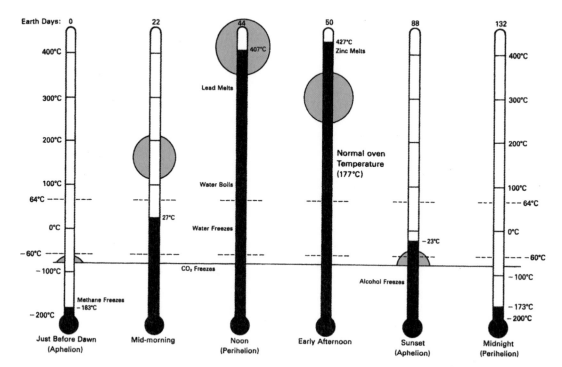

Figure 3.5. Mercury experiences the most extreme temperature range of any planet or satellite in the Solar System. This diagram shows the change in temperature of Mercury's surface during a solar day. Also shown are the freezing and melting points of several substances. The temperatures between −60°C and 64°C are the range of temperatures that can occur on Earth; −60°C is the coldest temperature ever recorded and 64°C is the hottest (from Strom, 1987).

610°C or 1,130°F. No other planet or satellite experiences such a wide range in temperature (Figure 3.5). The reasons for such enormous temperature extremes are the intense solar radiation, the lack of a dense insulating atmosphere, and the length of a Mercurian day where sunrise to sunset lasts nearly three Earth months. This gives the surface a long time to heat up. But Mercury's nights are just as long so the surface has a long period to cool.

3.2   MERCURY AND RELATIVITY

With time, Mercury's elliptical orbit is pulled around the Sun so that the perihelion point changes its position in space (Figure 3.6). The force that moves the orbit is mostly the gravitational attraction on Mercury by the other planets. In about a century, the perihelion point appears to move around the Sun by approximately 5,600 arcsec (less than 2°). When the effects of the Earth's precession (the

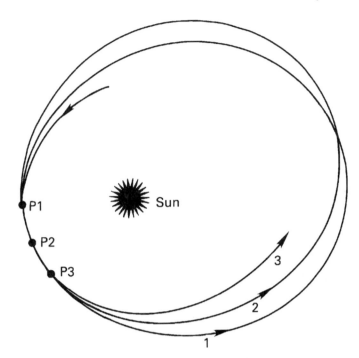

Figure 3.6. Mercury's perihelion point advances about 5,600 arcsec each century. In this diagram, the time taken for perihelion to advance from P1 to P3 (20°) is more than 1,280 years, or about 5,340 orbits. The perihelion advances 43 arcsec per century more than can be accounted for by perturbations by other planets. This discrepancy can be explained by Einstein's general theory of relativity, because at perihelion Mercury is travelling through space that is more warped by the Sun than at aphelion (from Strom, 1987).

changing orientation of the Earth's axis in space) and the gravitational pull of the other planets are subtracted from the 5,600 arcsec, there remains 43 arcsec per century unexplained by Newtonian gravitational theory.

Einstein's general theory of relativity predicts that the Sun's mass – 745 times the total mass of the planet – warps the nearby space. Since Mercury's orbit is so elliptical, at perihelion it is traveling in space that is more warped than at aphelion. The predicted advance in Mercury's perihelion by moving in this warped space is 43 arcsec per century, in agreement with the measured value. This agreement between Mercury's observed perihelion advance and that predicted by relativity theory has been cited as one proof of the validity of Einstein's general theory of relativity.

4

Mercury's size, mass, and density

4.1 MERCURY'S SIZE

Mercury is the smallest planet in the Solar System (4878 km diameter). Even three outer planet satellites are equal to or larger than Mercury – Callisto (4818 km diameter) is a satellite of Jupiter and almost the same size as Mercury, Ganymede (another satellite of Jupiter) at 5,468 km diameter is significantly larger and the Saturnian satellite Titan (5,150 km diameter) is also larger. Mercury is only 4,878 km in diameter, or about one-third the diameter of Earth (Figure 4.1). Its volume is only about 6% that of the Earth, so it would take almost 18 Mercurys to make one Earth (Figure 4.2).

4.2 MASS AND SURFACE GRAVITY

Although Mercury is small, it is very massive for its size (3.3×10^{23} kg). Therefore, Mercury has almost the same surface gravity as the larger planet Mars. A planet's surface gravity is a measure of how fast an object is accelerated by its gravity, and is usually measured in cm/sec^2. On Earth, a dropped object will increase its speed by 980 cm/sec per second. The amount by which an object accelerates is determined by the mass and radius of the planet. Although Mercury is 30% smaller than Mars, the combination of its large mass and small size results in a similar surface gravity (370 cm/sec^2). For the same reason, Mercury's gravity field is more than twice as great as the Moon's, although its size is only 40% larger. Because it contains so much mass in relation to its size Mercury possesses an unusually high density, only slightly exceeded by the density of Earth.

Figure 4.1. Mercury is about one-third the diameter of Earth.

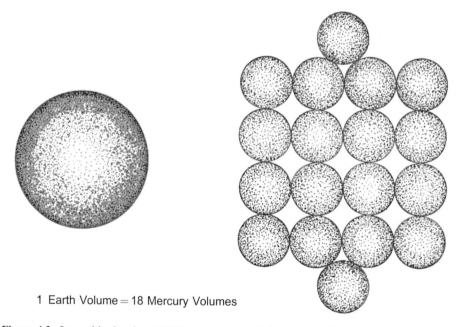

1 Earth Volume = 18 Mercury Volumes

Figure 4.2. It would take about 18 Mercurys to equal the volume of Earth (from Strom, 1987).

4.3   DENSITY

The density of a planet or satellite reveals something about its gross composition and internal constitution. Density is determined by dividing the mass by the volume. It indicates how much mass is contained in a unit of volume. Density is measured in g/cm^3 or kg/m^3. We will use g/cm^3 throughout this book.

4.3.1   What are the important issues relating to density?

Planets and satellites are composed of material made up of elements with different atomic masses. Iron is a heavy element with a large atomic mass, while hydrogen and oxygen are relatively light elements with small atomic masses. Therefore, a cubic centimeter of iron has a much greater density than a cubic centimeter of water; iron has a density of $7 g/cm^3$, and water has a density of $1 g/cm^3$ (Figure 4.4). Igneous rocks are composed of various silicate and oxide minerals. They have densities ranging from about 2.6 to $3.3 g/cm^3$ depending on their composition. For example, the Hawaiian Islands are composed mostly of a volcanic rock called basalt that is rich in minerals containing iron. Basalt, therefore, has a relatively high density of about $3.0 g/cm^3$. Granite, on the other hand, is poor in iron-bearing minerals and thus has a relatively low density of about $2.7 g/cm^3$.

Density can be increased by applying pressure. As a substance is compressed, the atoms of that substance are forced into a smaller volume, and the density increases (Figure 4.3). For instance, a cubic centimeter of basalt compressed to half its volume

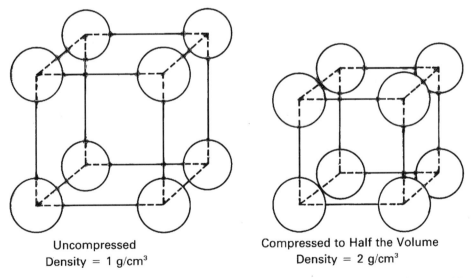

Uncompressed	Compressed to Half the Volume
Density = 1 g/cm³	Density = 2 g/cm³

Figure 4.3. A large amount of pressure can cause the density of a substance to increase. If a material is compressed to half its volume, it will still have the same mass but its density will be twice that of its original uncompressed state (from Strom, 1987).

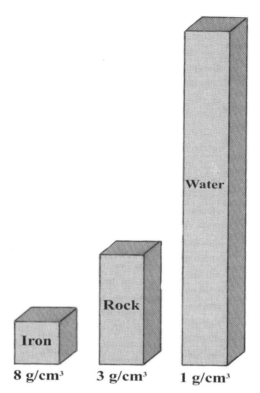

Figure 4.4. Each of these substances possesses the same amount of mass but different densities and therefore volumes (from Strom, 1987).

will contain the same number of atoms but now occupying only half the space. Thus, a cubic centimeter of this compressed basalt will now have a density of $6\,g/cm^3$ instead of its uncompressed density of $3\,g/cm^3$. Of course, extreme pressures, found only in the interior of relatively large planets, are capable of causing increases in density.

The Earth as a whole has a density of $5.5\ g/cm^3$, which is about halfway between the average density of rocks and iron. This density represents an average density of the iron core, consisting of 16% of the Earth's volume, and the rocky mantle and crust, which make up 84%. Because the Earth is so large, pressures in the interior are extremely high. These high pressures compress the atoms of a material into a smaller volume so that they are more closely packed together than for the same material nearer the surface. Thus, a rock or metal deep in the interior of Earth will have a higher density than the same rock or metal found near the surface. When these density or phase changes are taken into account and corrected for the Earth's pressure gradient, the Earth's uncompressed average density is only about $4.0\,g/cm^3$ (Figure 4.5).

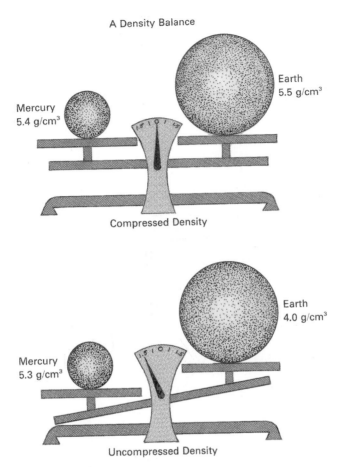

Figure 4.5. The compressed densities of the Earth and Mercury differ by only about 0.1 g/cm^3, Earth's density being slightly greater. The difference between the uncompressed densities, however, is about 1.3 g/cm^3, with Mercury's significantly greater than Earth's (from Strom, 1987).

4.3.2 Mercury has the greatest uncompressed density of any planet

Mercury has a density of 5.4 g/cm^3, which is comparable to that of Earth and Venus (5.2 g/cm^3) but much larger than that of the Moon (3.3 g/cm^3) or Mars (3.9 g/cm^3). However, Mercury is much smaller than Earth and, therefore, pressures in its interior are considerably less than those in Earth's interior. When this factor is taken into account, Mercury's uncompressed average density is still a high 5.3 g/cm^3, which is much larger than Earth's uncompressed density. This must mean that Mercury is composed to a large extent of heavy elements. Iron is the most abundant heavy element in the Solar System and is an important constituent of meteorites and terrestrial planet rocks. Seismic data for Earth also indicate that

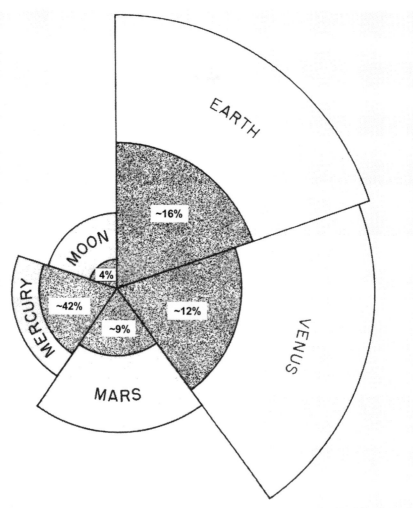

Figure 4.6. Relative sizes of the Moon and terrestrial planets and their cores. The % volume of the cores is also shown (from Strom, 1987).

Earth's core is mostly iron. Scientists strongly suspect, therefore, that iron is the principal heavy element responsible for Mercury's high density. From this high density we can infer that the planet is composed of about 70% by weight of metallic iron and only about 30% by weight of rocky material. Mercury thus contains more than twice as much iron per unit volume as any other planet or satellite in the Solar System. This iron is probably concentrated into a core like Earth's, but its size is enormous compared to the diameter of Mercury. The diameter of the core is about 75% of Mercury's total diameter and constitutes about 42% of its volume. In contrast, Earth's iron core is 54% of the total

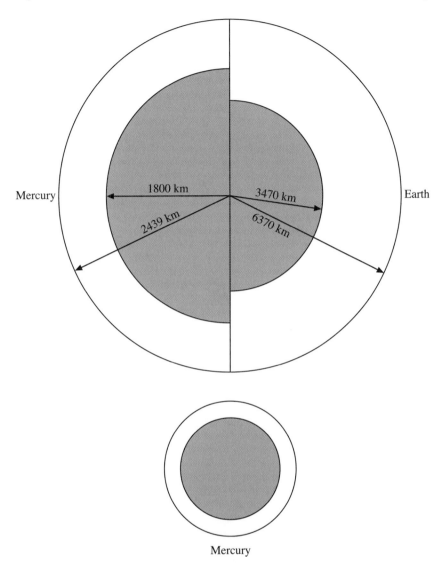

Figure 4.7. This diagram shows Mercury and Earth scaled to the same diameter. It illustrates how large Mercury's core is compared to the planet's total size. Mercury's actual size in relation to the Earth's is indicated by the smaller disc at the bottom (from Strom, 1987).

diameter but constitutes only 16% of the volume (Figures 4.6, 4.7 and [4.8, see colour plate section]). How Mercury obtained such an enormous iron core has important implications for Mercury's origin, which in turn sheds light on the origin of all the terrestrial planets. This will be discussed in the Chapter 12.

5

Mercury's magnetic field and internal constitution

5.1 A MERE THIRTY MINUTES

Mariner 10 first encountered Mercury's magnetosphere on 29 March, 1974 at 2037 UT as it approached Mercury. It was at a distance of 1.9 radii from the planet surface. Magnetic field and charged particle instrumentation made measurements for only short periods of the first and third flybys of the planet, but investigators were able to piece together a picture of the magnetic field environment at Mercury based on analogy with that of Earth's magnetic field and particle environment. However, because Mercury's atmosphere is very thin and the magnetosphere is very small, it probably lacks the *ionosphere* and trapped *radiation zones* of Earth's *magnetosphere*. Signals in the ∼30 minutes of data (∼17 minutes during the Mercury 1 equatorial pass and ∼13 minutes during the Mercury 3 high-latitude pass) provide all we know about Mercury's magnetic field, magnetosphere, and particle environment.

5.1.1 A significant magnetic field discovered

The measured magnetic field is strong enough to present an obstacle to the solar wind, which streams towards Mercury in a manner similar to that which occurs at Earth. As the spacecraft approached the planet, it first encountered a sudden jump in the magnetic field associated with the *bow shock*. In addition, signals measured by the instruments indicated entry into and exit from a *magnetopause* surrounding a magnetospheric cavity. The magnetospheric cavity is estimated to be a factor of about 20 times smaller than the Earth's magnetic cavity and would fit in an Earth-sized sphere of about 12,000 kilometers in diameter. The magnetic polarity is the same as that of the Earth with respect to the direction of the angular momentum vector associated with the planet's rotation. The magnetometer experiment measured an increase in the magnetic field as the spacecraft approached the

Table 5.1. Third encounter magnetometer results.

Event	Time of observation Pacific Daylight Time (PDT) hr:min	
	Predicted	Actual
Cross bow shock	3:31 ± 02	3:31
Cross magnetopause	3:39 ± 01	3:39
Maximum field*	3:49 ± 01	3:49
Recross magnetopause	3:54 ± 01	3:56
Recross bow shock	3:58 ± 02	5:59

* Predicted field strength was 200–500 nT; actual strength was 400 nT.

planet. The *interplanetary magnetic field* (IMF) typically has a strength of about 6 nT (Tesla – units to measure magnetic intensity) at Earth and about 25 nT in the vicinity of Mercury, but at closest approach to Mercury the magnetic field strength reached 100 nT. If the rate of increase continued to the surface, Mercury would have a ground-level magnetic field of about 200 to 500 nT. Although this strength is only about 1 percent of Earth's field, it is adequate to deflect or stand off the solar wind and produce the bow shock observed by plasma and charged particle experiments. Table 5.1 compares the actual measurements to a first encounter predicted model used to infer the shape and strength of the magnetic field at Mercury.

5.1.2 Comparison to Earth's magnetic field

On Earth, the configuration of the magnetic field, the magnetosphere, and surrounding interplanetary medium are well-known from extensive ground and orbital spacecraft measurements. The Earth's field is believed to be generated by a dynamo and changes by up to 100 nT per year at the surface in some places. The geologic records show that it changes polarity every 500,000 years or so. We know from seismic data that the Earth has a fluid outer core and the heat flux of the interior has more than enough energy to support a dynamo. We use these facts to build a model for the same concepts on Mercury even though there is great uncertainty because of the scant amount of data. If we assume that our interpretation of the data is correct, and Mercury does indeed have a smaller, but similar magnetic field to that of Earth, then we can illustrate these concepts in Figure 5.1.

The solid body of Mercury fills a greater portion of its magnetosphere than does the Earth. The cusp regions, where the field lines intersect the surface of the planet are probably at lower latitudes on Mercury than on Earth, and the magnetosphere probably fluctuates more rapidly and more often than on Earth as a result of more perturbations from the solar wind – *flares, coronal mass ejections,* and other solar related phenomena.

The dynamic disturbances of particles and fields measured *downstream* from Mercury appeared similar to substorms in the Earth's *magnetotail* with its

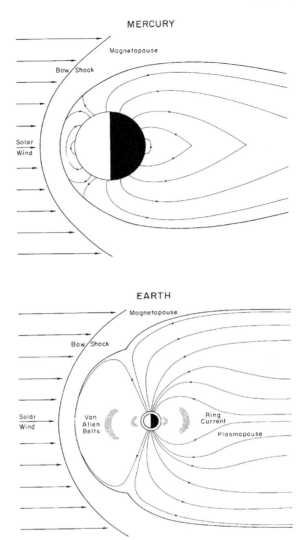

Figure 5.1. A comparison of models for the magnetic fields of Mercury and the Earth. Mercury's is assumed dipolar. Mercury fills a smaller portion of the magnetic cavity in the solar wind because its field is about 1% that of the Earth (from R. G. Strom, 1987).

associated *plasma sheet* and *cross-tail currents*. A color illustration of Mercury's magnetosphere and its interaction with the solar wind is shown in Figure 5.2. (color plate section).

5.1.3   Mercury's magnetic field could be remanent

It is possible that the interpretation of the *Mariner 10* particles and fields instruments was biased by comparison to the known field of the Earth. However, this does not seem likely as the evidence for the current interpretation following the three *Mariner 10* encounters was convincing. Nevertheless, it may turn out that Mercury's fields are largely *remanent fields* left in the crust from a much earlier period. We have good examples of such remanent fields in our Solar System on the Moon, Mars, and Earth.

5.2 REMANENT MAGNETIC FIELDS ON THE MOON

The remanent magnetic fields at the lunar surface have been a source of intense interest and debate since their discovery with instrumentation on the *Apollo* landers and orbiters during the 1960s and 1970s. Localized magnetic fields up to several tens of nT were found at orbital altitudes above the antipodes of the Imbrium and Crisium basins, and there fields are associated with bright albedo features scattered across the lunar surface. More recent observations from instruments on the *Lunar Prospector* spacecraft found some evidence for surface fields as high as 200 nT. Suggestions for their sources include: magnetized ejecta from deep impacts, crustal material shocked in the presence of an existing field (shock magnetization), or plasma cloud magnetic enhancement of surface regolith concomitant with large impacts. For the third hypothesis, the suggestion is that reduced Fe in the regolith is magnetized directly by the enhanced field of the impacting plasma.

Associated with some of the greatest magnetic anomalies are surface regions of considerably higher *albedo* than the surrounding soils. One of the best known examples of intense localized magnetic fields is Reiner Gamma in western Oceanus Procellarum shown in Figure 5.3. This region consists of a bright magnetic region surrounded by darker soils outside the magnetic region. It has been suggested that the magnetic fields in these regions have protected the surface from darkening by shielding soils from the solar wind and charged particle sputtering.

5.3 MARS' REMANENT MAGNETIC FIELDS

In 1998, the magnetometer/electron reflectometer (MAG/ER) experiment on *Mars Global Surveyor* (*MGS*) made a startling discovery of intrinsic intense magnetization mainly confined to the heavily cratered, ancient southern highlands. These remanent magnetic fields reach a maximum strength of ∼220 nT at the *MGS* mapping altitude of 370–438 km. The fields are thought to be "frozen" into the rocks and are consistent with a reversing dynamo halted in an earlier period when they were generated in the deep interior. This was a surprising discovery because previous orbiters had not detected strong magnetic fields on or around Mars' surface. Previous studies of the upper atmosphere had not detected a "stand-off" region between the upper atmosphere of Mars and the solar wind. Some of these remanent fields are quite strong, about the strength of the magnetic field measured on Mercury. At Mars

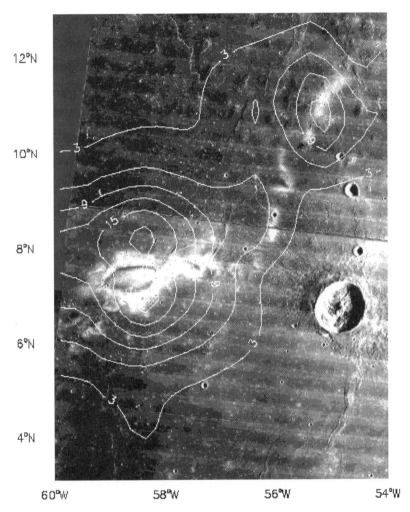

Figure 5.3. The remanent magnetic fields on the Moon were measured with the magnetometer and electron reflectometer on *Lunar Prospector*, an orbiter sent to the Moon in 1998. Contours of field strength in nT are plotted over an image of the Reiner Gamma region in western Oceanus Procellarum obtained by *Lunar Orbiter 4* in 1967. The 30-km-diameter crater Reiner is at the lower right (courtesy of Lon Hood, University of Arizona).

many orbits of MGS were able to map the magnetic regions fully and determine their strength and extent over the surface as shown in Figure 5.4.

Mars has a highly oxidized surface, although at this time the exact details of the composition of the dust and rocky competent layers are not known. However, much of the planet is covered by basalt and its weathering products. Regions of hematite have also been discovered on the surface by identifying its spectral signatures in data from the Thermal Emission Spectrometer (TES) on the *MGS*. Thus, plausible material for the source of the strong remanent fields are titanomagnetite or

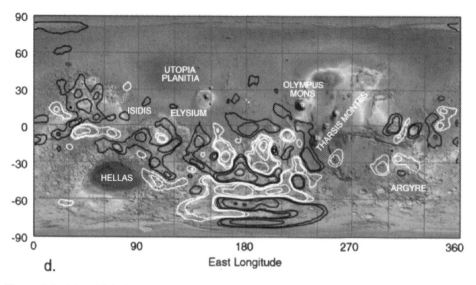

d.

Figure 5.4. *Mars Global Surveyor* discovered magnetic fields in Mars' crustal materials. The strength of the radial magnetic fields vary from $B_r = -160$ to $+160\,$nT. Contour lines on the map show the locations and strengths of the magnetic fields B from -10 to $200\,$nT, with black low and white high (courtesy of *Mars Global Surveyor* particles and fields scientific team).

titanohematite. Iron sulfides (pyrrhotite) have also been suggested as plausible remanence carriers in the deep crust of Mars. There are other possibilities but the exact composition remains unknown.

5.4   A NEW LOOK AT OLD DATA

Recent spacecraft exploration of Mars has resulted in new studies and insights regarding the remanent magnetic fields. One explanation for the distribution of the newly discovered remanent magnetic fields in the southern highlands is that impact shock demagnetization occurred during the very large Hellas and Argyre impacts. These impacts would have raised the temperature of the crust in and near the impact areas to values above the Curie point and demagnetized the rocks.

Now, because of the new discoveries on Mars, scientists are taking a new look at the *Mariner 10* data from both the particles and fields and magnetometer instruments. Some scientists are seriously examining the *Mariner 10* data to see if it too could be interpreted in terms of irregularly distributed strong remanent fields on Mercury's surface rather than the dipolar configuration that scientists debated when the data were originally interpreted in the 1970s. Knowing that their conclusions rested on a mere thirty minutes of data leaves considerable uncertainty in some researchers minds. If Mercury's magnetic field is remanent, then it will be the first dipolar remanent field ever discovered. Since there is no known way to take more measurements from Earth, it is crucial to go back to Mercury and measure its magnetic environment from orbit.

5.5   INTERIOR STRUCTURE AND CONSTITUTION

Our current understanding of planetary dipolar magnetic fields is that they are generated by electrical currents induced by dynamo action in a thermally convecting, differentially rotating, liquid metallic core. Most scientists believe such a mechanism causes the Earth's dipolar magnetic field. For a large planet like the Earth, it is not unlikely for a liquid core to remain even four billion years after planetary formation. Seismic data from many Earth-based seismograph networks confirm the existence of the liquid outer core and a solid inner core.

Mercury, however, is small compared to Earth. It seems likely that its core would have cooled and solidified during the past four billion years. Many theoretical studies have attempted to explain how a small planet like Mercury could keep a liquid outer core and generate the measured magnetic field. Some models which include light elements in the core like sulfur, oxygen, or silicon show that a liquid outer core at this time in geologic history is possible. But we are terribly hampered by lack of both chemical and dynamic data. Detailed knowledge of the elemental surface composition and of the gravitational moments of the planets are required before further modeling will be useful. Many more details relating to the interior structure and composition are discussed in Chapter 12.

5.5.1   What if there is an active dynamo?

If the magnetic field is a presently active dipole and Mercury presently has an outer fluid core, physical mechanisms must maintain high core temperatures up to the present time. Proposed means of facilitating this are: (1) provide more internal heat by enriching the core in the radioactive elements uranium and thorium; (2) retain the heat longer by reducing the thermal diffusivity of the mantle; or (3) add some light alloying element to lower the melting point of iron. The addition of a light alloying element is considered to be the most likely cause. Although oxygen is such an element, it is not sufficiently soluble in iron at Mercury's low internal pressures. Metallic silicon has been suggested, but sulfur is considered to be the most likely candidate. For a sulfur abundance in the core of less than 0.2%, the entire core should be solidified at the present time, and for an abundance of 7% the core should be entirely fluid at the present time. Therefore, if the outer core is presently molten and sulfur is the alloying agent, then Mercury probably contains between 0.2–7% sulfur in the core.

5.5.2   What if there isn't?

If the magnetic field is a remanent field it permits the possibility of a completely solid core at present. Thus the sulfur content could be less than 0.2%. For a sulfur content of 0.2%, the core takes almost the age of the Solar System (4.6 billion years) to solidify. With less it would solidify much faster, but it would still take well over a billion years to completely solidify because of its very large size. However, if Mercury's field is remanent then it requires a thick layer of abundant magnetic minerals (perhaps more than 30 km thick) which may be unrealistic from a geochemical and petrological standpoint.

5.5.3   Relevance of Mercury's surface composition to the magnetic field question

Magnetic minerals responsible for crustal magnetic remanence must contain iron in a form that can acquire and preserve magnetic fields. The ability of rocks and minerals to maintain magnetic fields over long time periods is called *remanence*. One of the more common minerals, and one that may be responsible for remanent fields on the Moon, is metallic iron. On Mars a possible candidate is titanohematite, a solid solution mineral of Fe_2O_3 and $FeTiO_3$. Titanomagnetite, a solid solution of two end-member minerals Fe_3O_4 and Fe_2TiO_4 is another possibility. A solid solution can exist with varying amounts of each of the two end member minerals comprising its makeup. Because the surface composition of Mercury appears to be low or devoid of FeO-bearing materials, oxidized iron-bearing silicates would have to be buried for the remanent field scenario to be possible if it were caused by remanence of titanomagnetite or titanohematite. Alternatively, with Mercury's other similarities to the Moon, it could be that the remanent fields, if they exist, would be associated with metallic iron.

However, another mineral that has remanence is pyrrhotite, Fe_7S_8. Iron sulfides have been postulated as possible components of Mercury's regolith as a possible source of sulfur in the atmosphere and deposits in high-latitude cold craters where the high coherent backscatter of radar signals has been observed. There will be much more about the high-latitude radar backscatter images and their possible causes in Chapter 7. Until measurements positively identify sulfur in some form on Mercury, all of these suggestions remain speculative.

5.6   SPACE WEATHER AND SPACE WEATHERING ON MERCURY

The Sun is constantly emitting particles, as well as light, into the Solar System. Ions, mostly protons, and electrons stream from the Sun and move throughout the Solar System dragging solar magnetic fields with them. The average geometry of these streaming fields is a spiral ever increasing in distance from the Sun. The particles emanating from the Sun make up the *solar wind*. The spiraling magnetic field (*Parker spiral*) is called the interplanetary magnetic field (IMF).

5.6.1   Space weather

Fluctuations and rapid changes in the "average state" of the solar wind and IMF are called *space weather*. At times of high solar activity, large flares leap out from the Sun carrying unusually large numbers of charged particles and distorting the normal positions of the magnetic field in the solar wind. Sometimes the x-ray and ultraviolet (UV) flux from the Sun increases drastically for short periods of time. The corona of the Sun may emit ejections of atoms and ions into the Solar System. Collectively, these phenomena of rapid changes in sunlight, solar particles, and fields are called space weather. On Earth we are familiar with satellite-to-Earth communications being occasionally interrupted during periods of unusual *space weather*. On Mercury the effects of space weather are even stronger than on Earth because it is closer to the Sun and the intrinsic magnetic field of the planet is smaller than that of Earth.

An intrinsic magnetic field like that of the Earth can hold off the solar wind as shown in Figure 5.2. Although charged particles and fields can enter into the Earth's magnetic field near the magnetic poles, when they do, they encounter the Earth's atmosphere. Interaction of space weather and the Earth s upper atmosphere and ionosphere result in high-altitude chemical changes and auroral displays. Earth's surface is largely sheltered from the effects of protons, other ions, and fields from the Sun.

Ions traveling parallel to the magnetic field lines near Mercury may come from the Sun or from Mercury's surface or atmosphere. Most ions are H^+ or He^+, but we know that there are Na^+, K^+, and Ca^+ in the vicinity of Mercury because the neutral counterparts of these ions have been discovered in Mercury's exosphere (there will be more about this in Chapter 6). There may be S^+, OH^+, O^+, and other ions from *photoionization* of neutrals delivered to Mercury s atmosphere from volatilization of meteoritic and surface materials. These ions may be recycled by induced electric fields near Mercury and redirected toward Mercury's surface to be neutralized and stored in cool places, or, they may be swept away from the planet to join other ions in the IMF.

5.6.2   Space weathering

Space weathering is distinct from space weather. Space weathering is what happens to surfaces of asteroids, the Moon, and Mercury as they undergo modification from sunlight, meteoritic bombardment, charged particle sputtering, and assaults from a myriad of other physical processes in space.

Modeling of Mercury's magnetic field, based on *Mariner 10* data illustrated in Table 5.1. showed that Mercury s magnetosphere could push against, and hold off, the solar wind from Mercury's surface at low- and mid-latitudes. But because the field at Mercury is relatively small, and there is only a very thin atmosphere on Mercury, regions on Mercury's surface are subject to ion and electron bombardment. Also the surface is directly subjected to the effects of sunlight at all wavelengths because there is no thick attenuating atmosphere like on Earth or Venus to diminish short wavelength (extreme ultraviolet (EUV) and UV) light. Thus Mercury's surface undergoes space weathering. This will discussed in more detail in later chapters.

6

Mercury's surface-bounded exosphere

6.1 LIGHT EMITTING GASES

In the space surrounding Mercury atoms rarely collide, and, therefore, it is properly termed an exosphere rather than an atmosphere where the atoms are in constant collision. For decades carbon dioxide (CO_2) was thought to be the most likely gas in Mercury's exosphere because it had been discovered on Mars and Venus. Several attempts with ground-based spectrographs were made to find CO_2 absorption bands during the 1960s and 1970s but no absorptions were found. This was not too surprising because Mercury is a small planet with relatively low gravity and it was not likely that any gases ejected into the exosphere early in its history would remain bound to the planet. Nevertheless, *Mariner 10* was equipped with instrumentation to search for potential atoms and molecules in the exosphere, and made the first discoveries during the three flybys of Mercury in 1974 and 1975 as described in Chapter 2.

6.1.1 Hydrogen, helium, and oxygen

In *Mariner 10* airglow experiments, limited wavelength bands were predetermined based upon optical requirements and expectations of what might exist in an entirely unknown exosphere. Hydrogen (H), helium (He), and oxygen (O) were identified with the airglow polychromater that had 10 wavelength channels to search for light emissions. Upper limits on the abundance of neon (Ne), argon (Ar), and carbon (C) were also obtained.

6.1.2 A serindipitous discovery

The amount of light coming from the Sun's photosphere varies with wavelength because atoms in the photosphere absorb sunlight coming from deeper in the Sun.

Table 6.1. Mercury's exospheric species.

Constituent	Vertical (zenith) column abundance (atoms per cm^2)
Hydrogen (H)	$\sim 5 \times 10^{10}$
Helium (He)	$\sim 2 \times 10^{13}$
Oxygen (O)	$\sim 7 \times 10^{12}$
Sodium (Na)	$\sim 2 \times 10^{12}$
Potassium (K)	$\sim 1 \times 10^{10}$
Calcium (Ca)	$\sim 1 \times 10^{7}$

Such absorption lines are called *Fraunhofer lines*. When sunlight is reflected off the surface of the Moon, or Mercury, or any other Solar System body, the Fraunhofer lines appear as distinct regions of less sunlight than the average. The average is called the *continuum*. In 1985, planetary astronomers Andrew Potter and Thomas Morgan were studying a phenomenon called the Ring Effect (the infilling of solar Fraunhofer lines in the reflected continuum from the lunar surface). For comparison, they shifted their view to Mercury and observed significant emission lines high above the continuum at 5890 and 5896 Å. They had discovered Sodium (Na), the first new species in Mercury's exosphere since *Mariner 10* in 1974. This discovery renewed ground-based search efforts to find more atmospheric components. Discovery of potassium (K), calcium (Ca) and an upper limit on lithium (Li) have followed.

Table 6.1 shows all known constituents of Mercury's exosphere and their approximate abundances. The abundances of Na and K are known to vary by a factor of 10 or more from one measurement to the next or even at different locations above the planet during the same observation period. Probably the abundance of other species also varies but we do not have enough observations to be certain. Zenith column abundance refers to the column of atoms extending in a vertical direction from the ground to the top of the exosphere with the column foot print being a square with 1 cm length sides or 1 square centimeter. Because measurements do not measure the vertical column directly, the zenith column abundance depends upon the model used to calculate its value. The model is based on actual measurements. The Earth's atmosphere has $\sim 2 \times 10^{18}$ molecules per square centimeter.

6.1.3   Sunlight interacting with matter

All of the known species have been discovered by measuring the emission of solar *photons* from the atoms of Mercury's exosphere after interaction with the electrons in these atoms. Each type of atom has one or more wavelengths at which it absorbs and reemits sunlight. The particular wavelength is determined by the electronic structure of the atom. Light absorbed and emitted at the same wavelength for a particular species, it is called *resonance scattering*. The process is similar to, but distinct from, fluorescence where light is absorbed at one wavelength and emitted at another.

Light is measured by spectrographs as it is reflected off the surface of Mercury and as it is scattered by atoms in Mercury's exosphere (Figure 6.1). The measurement results in a spectrum in which the continuum, the depressions of the Fraunhofer lines, and the emission lines are present. The amount of photons emitted by the atoms in the exosphere is measured in units of Rayleighs (R). Sodium emits about 1 million photons per second in a $1\,cm^2$ column. This amount is called a mega-Rayleigh (MR).

A slight offset in wavelength from the peak of the emission line and the depth of the absorption lines is a result of Mercury moving relative to the Earth at the time of measurement, and, thus, its emitted light is shifted with respect to the wavelength of the solar feature. This shift is called a *Doppler shift*.

6.1.4   Exospheric pressure?

The pressure of the known exosphere is a few times 10^{-12} bar (b). This is $\sim1/1,000,000,000,000$ the pressure of Earth's 1 bar atmosphere. Mercury's atmospheric atoms do not collide appreciably with one another, only with the planet surface. For this reason it is called a surface-bounded exosphere. Mercury's exosphere has multiple speed distributions with some, and perhaps all, species having multiple speed components that result from differing source, release, and recycling mechanisms.

6.1.5   New discoveries are likely

It is likely that NASA's *MESSENGER* mission (see Chapter 13) will make new exospheric discoveries with its atmospheric spectrograph (MASCS), because it analyzes the region from 1150–6000 Å with the ability to observe very faint resonance emissions above Mercury's surface.

Two gases of particular interest are sulfur (S) and hydroxide (OH). This is because ground-based very high-resolution radar imaging of Mercury's surface has revealed deposits at latitudes greater than $\sim70°N$ and S. The deposits are in permanently shadowed craters, thus remaining very cold over geologic periods of time (billions of years). These deposits have radar reflection and polarization characteristics that are similar to deposits of water ice on Mars and the *Galilean satellites*. However, this similarity could be caused by other substances, such as S, that are very transparent to radar signals at 7 and 35 cm (the wavelengths used in many radar studies). Other possibilities must also exist but are not presently known.

If the substance stored in the shadowed craters is water ice then it is possible that there will be a thin water vapor signature above the region. When photons break apart the water vapor molecule, disassociation of its two neutral components, OH and H occurs. While the H_2O cannot be observed directly, the H can be detected by a neutron mass spectrometer and the OH by an ultraviolet spectrometer. If these substances are both found by *MESSENGER*, or some other spacecraft like *Bepi Colombo*, then it will be excellent evidence that the stored volatiles are water ice.

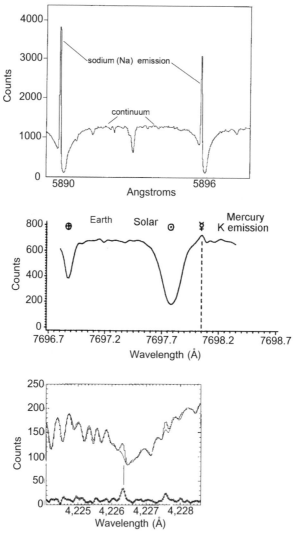

Figure 6.1. Actual spectra from Mercury's exosphere of sodium (Na) (top), potassium (K) (middle), and calcium (Ca) (bottom). In the spectrum of Na the sharp emission peaks at 5890 Å and 5896 Å are to the left of the Solar Fraunhofer absorptions. In this case there is a *blue-shift*. In the middle spectrum the small bump identified with the astronomical symbol for Mercury is the K emission line. For the K measurement there was a *red-shift*. A dotted vertical line below the emission peak guides the eye to the wavelength scale of the figure. The two dips in the spectrum are from Earth's atmospheric absorptions (marked with the Earth symbol ⊕) and an absorption caused by K in the Sun's atmosphere – the Solar Fraunhofer line (marked with the Solar symbol ☉). Much more complicated is the spectrum from the wavelength region where Ca scatters light in Mercury's exosphere (bottom spectrum). The Fraunhofer absorption lines of Ca in the solar spectrum are wide and the many other absorptions are from other molecular absorptions in the Earth's atmosphere. The emission peak from Ca is denoted by a vertical line above (after continuum removal) and below (actual spectrum).

If, however, emissions at 1636 Å (one of the wavelengths at which atomic S has its resonance scattering emissions) are found by the ultraviolet spectrometer, then it will be good evidence for S being the substance in the shadowed craters. Another possibility is that ordinary silicates that are low in iron may also give a radar back-scatter signal when they are very cold. Thus, the radar bright spots may not be volatiles at all! Imaging, along with X-ray and γ-ray measurements, and measure-ments by the surface and atmospheric spectrographs may reveal that the mysterious radar signal is caused by some other, as yet unknown, substance.

6.2   EXOSPHERIC ATOMS AND MULTIPLE SPEED COMPONENTS

6.2.1   Collisions only with the surface

Because atoms in the exosphere do not collide with one another, the speed and direction of their motion is determined by the release mechanism from the surface, and the arrangement and character of the surface grains. Much effort has been spent to understand these physical issues. Some details are gleaned from measurements of the light emitted from the atoms and relevant physical laws. For H and He at least two populations, high and low temperature groups, were discovered as a result of modeling density and height profiles measured above the planet's surface by instru-ments on *Mariner 10*. Daylight and darkness each last about 88 Earth days on Mercury, and the surface, therefore, reaches extremes of temperature because the exosphere is too thin to modify the surface temperature as happens on Earth. The cold gas component results from atoms hopping from the night side and the smaller warmer component is formed on the day side. Helium extends from 3000–4000 km above the surface, and thus, was seen against the nightside of the planet and well off the sunlit side where solar photons were able to interact with He atoms.

6.2.2   Detection of Multiple Speed Components

For Na, detection of multiple speed components was possible because the width and shape of the resonant emission line measured with ground-based telescopes and spectrographs, could be modeled. The first such measurements of the Na emission line showed that the Na atoms had a most probable speed of ~600 m/s with an equivalent temperature of ~230°C. For comparison, the atoms and molecules in Earth's atmosphere have a most probable speed of about 350 m/s and at a tempera-ture of about −3°C. The shape of the Mercury Na emission line also showed that a hotter population of atoms, perhaps up to ~1000 K, could be present. Later meas-urements did indeed exhibit characteristics of much warmer temperatures and higher speed atoms; different speeds being observed at the equator and in the polar regions (ranging from ~330°–1230°C).

For K, there are no measurements to date which can be used to discriminate multiple speed components in Mercury's exosphere. It is likely however, to also have atoms travelling at many different speeds. Ca was observed above and beyond the southern hemisphere of Mercury using a spectrograph and large telescope (Keck I) on Mauna Kea, Hawaii. The emission lines exhibit characteristics of high speed and high temperature ($\sim$12,000°C) atoms. The equivalent most probable speed for Ca is $\sim$2.2 km/s. The speed at which an atom will escape Mercury's gravitational field if it is moving straight away from the center of the planet is called the *escape velocity*. The speed of Ca is thus considerably less than Mercury's escape velocity of 4.2 km/s. Thus, what Ca is present about the planet is still in the grip of the planet's gravitational force and not likely to be lost until it is ionized and then, having an electrically positive charge, it can be carried away in the electric fields of the solar wind.

6.2.3   Atoms in escape

Because of pressure from sunlight, some atoms are pushed far from Mercury's surface and beyond the gravitational force that keeps them bound to the planet. Therefore, some atoms escape from Mercury. On Earth, the escape velocity is 11 km/s which is more than twice that of Mercury. The amount of sunlight pressure varies considerably as Mercury moves around the Sun – both the distance to the Sun and the speed of Mercury's motion are changing.

Clever observations from Kitt Peak Observatory using the McMath Pierce solar telescope were able to image the escaping Na atoms at times of high *radiation pressure* on Mercury. One such image (Figure 6.2, see colour plate section) shows the coma of Na about the sunlit crescent and the tail streaming behind the planet. The observations were tailored towards measurement of the downstream profile of the tail along its axis. West is on the right, and south is at the top of the figure. The position of Mercury and the portion of the illuminated crescent is shown with the Na coma in front, above and below. The Na tail is seen streaming behind.

6.3   ATMOSPHERIC SOURCE, RELEASE, AND RECYCLING PROCESSES

The multiple speed distributions of atoms in the exosphere indicate that there are multiple physical processes at work to provide new atoms and to release recycled ones back to the exosphere. New sources are materials from volatilized meteorites, atoms from freshly exposed rock or buried sources, and a small component of Na, H, and He that come from the Sun delivered as ions in the solar wind. Recycled atoms are those that have been in the exosphere and subsequently have been stored on the surface for a period of time before being re-released from the surface back to the exosphere. Sunlight and ions from the solar wind strike atoms on the surface and release them back into the exosphere. Evaporation, or thermal desorption, is a major release process for H, He, Na, and K.

6.3.1 Differentiating one source from another

There is a decade-long history attempting to decipher telescopic data in terms of the sources, release mechanisms, and recycling processes. Much of the data can be interpreted in more than one way. For example, images of Na emission in Mercury's exosphere often show bright regions that appear at different locations from one measurement to the next. Some scientists believe that the bright emission spots are seen over Na rich rocks on the surface that have been freshly exposed by meteoritic bombardment. Other scientists believe that the same emission spots are caused by ions sputtering Na off of the surface as they are directed along electric fields associated with the magnetic field of the planet (Figure 6.3, see colour plate section).

  More ground-based imaging and spectroscopy may be able to settle this issue. With systematic observations made with telescopes and imaging facilities with adequate pixel scale, it may be possible to determine if the bright spots always appear over fresh craters on the surface, or if they appear during times of active "space weather" from the Sun. Some examples of the ground-based observations we do have are shown in Figure 6.4 (see colour plate section). In this figure, the radar bright regions B (northern hemisphere) and A (southern hemisphere) are shown in all images. These regions appear bright in radar backscatter because the surfaces are rough at the tens of centimeters scale, probably because they are fresh impact craters and are surrounded by fresh blocks of crater ejecta. They are on the side of the planet not imaged by *Mariner 10*. Atmospheric emissions of Na and K have also been seen over the radar bright regions. In the recalibrated *Mariner 10* images the Kuiper Muraski crater complex appears as a very bright fresh region (represented by the letter K). The Caloris basin is also shown in the images of Figure 6.4 (represented by the letters CB). Enhanced Na and K (potassium) have been observed over Caloris Basin. These regions and atmospheric observations are shown in the nine panels of Figure 6.4 (see colour plate section).

6.3.2 Ions recycle back to the surface of Mercury in electric fields

As discussed in Chapter 5, the interplanetary magnetic field (IMF) is a source of ions in the vicinity of Mercury. Also ions created near Mercury by photolysis of atmospheric neutrals are controlled by the local electric fields. Such electric fields E, are created by charged particles moving with velocity V in the solar wind magnetic field B, as shown in Figure 6.5. It is possible that some of the Na and K enhancements are caused by emissions from neutral atoms that are created after Na^+ and K^+ strike Mercury's surface. Because the direction of fields in the IMF change periodically, this mechanism could account for some of the time variation observed in the Na and K emissions in the atmosphere.

6.3.3 Exosphere and surface link

Obviously the exosphere and surface are linked by the gas-surface interface – the place where collisions, storage, and release occurs. But, does the exospheric

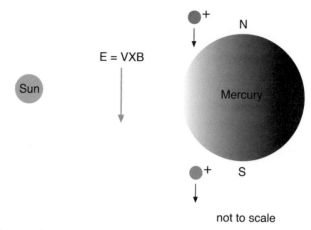

Figure 6.5. Na^+ and K^+ will interact in the electric fields in and around Mercury's magneto-sphere and the IMF. Such ions will eventually be swept away into the interplanetary medium or will impact on Mercury's surface, become neutralized and eventually be recycled to the atmosphere unless they find a permanent cold trap. Mercury's magnetic field is not considered in the diagram. See text for explanation of E, V, and B.

composition reflect, in a significant way, the surface composition? It is of great importance to know if Mercury has more Na and K than is expected from our understanding of how planets form. Also important is whether Mercury ever had water, and if so how much? The study of the exosphere may help us answer these questions.

7

General surface features and radar characteristics

7.1 IMAGING MERCURY

Mariner 10 photographed only about 45% of Mercury's surface. The image resolution ranges from about 2 km down to 100 m at a few locations. Images revealed a heavily cratered and wrinkled terrain with no obvious signs of volcanic constructs or plate tectonics. The first glimpse of the images was tremendously exciting because no detailed ground-based images of the planet existed, and maps and drawings were lacking in detail, and seldom agreed on what detail they did exhibit. The generalized statement was "Mercury looks a lot like the Moon." In fact, this generalized impression has lingered in the minds of people although new and surprising ground-based observations show it to be unlike the Moon in several important ways. The internal characteristics are totally unlike the Moon.

7.1.1 Photomosaics of one hemisphere

As *Mariner 10* approached Mercury on its first encounter, it imaged the half-lit hemisphere centered on the prime meridian, 0° longitude. As it departed, *Mariner 10* imaged the opposite half-lit hemisphere centered on the 180° meridian (Figure 7.1). On the second encounter, the spacecraft imaged the south polar region, joining the two sides previously imaged on the first encounter. On the third encounter *Mariner 10* concentrated on taking high-resolution images of the two hemispheres imaged on the first encounter. Unfortunately, by this time the tape recorder had failed and the spacecraft was so far from Earth that the signal was very weak. Full frame real time images would have been so noisy that they would have been of little use. Therefore, only $\frac{1}{4}$ frame images were transmitted. Although almost all of the hemisphere between 10 and 180° was imaged, much of it was seen at very at high Sun angles where terrain analysis is difficult to impossible. As a consequence only about 25% of the planet was viewed at Sun angles that were favorable to geologic studies.

(a)

Figure 7.1. Photomosaics of the incoming (a) and outgoing (b) sides of Mercury as viewed by *Mariner 10*. Most of the smooth plains are concentrated on the outgoing side.

Whether or not the major features seen on this part of Mercury are representative of the planet as a whole is not known and must await further exploration. The first two flybys imaged almost all of the hemisphere between 10 and 190° longitude. Only about 5% of the northern part of this hemisphere was not imaged. Part of the same surface imaged at different viewing angles between the first and second encounters provided some stereo coverage which was useful in geologically mapping the surface.

7.2   MAJOR SURFACE FEATURES

The *Mariner 10* images revealed a heavily cratered planet with some large patches of smooth plains. There were also sinuous ridges and a peculiar broken-up terrain on the incoming side. The largest feature viewed by *Mariner 10* was the half illuminated Caloris basin. This basin is about 1300 km in diameter and has a peculiar fractured and ridged floor. In general, Mercury looked similar to the Moon (Figure 7.1).

7.2.1   Just another moon?

Up to this point it may seem that Mercury and the Moon have a lot in common. Some major differences and similarities have been pointed out, but many people look at Mercury and see it as the Moon's twin with a similar history. This perception could not be farther from the truth. Mercury is distinctly different from the Moon in many fundamental ways. Its similarities, although important, are few compared to its differences (Table 7.1).

The main similarity between Mercury and the Moon is that both display heavily cratered highlands that are the result of the period of late heavy bombard-

Table 7.1. Comparisons between Mercury and the Moon.

Similarities to the Moon

Heavily cratered surface
Smooth plains associated with impact craters and basins
Regolith (impact produced surface layer)

Differences from the Moon
Large iron core ~75% of the diameter
Relatively strong magnetic field
Exosphere dominated by sodium (Na) and potassium (K)
Large areas of intercrater plains (the major terrain type)
Comparable geologic units are brighter
Widespread distribution of thrust faults
Unique impact basin floor structure (Caloris basin)
Unique radar feature
Very strong radar backscatter from polar deposits
Origin

ment which ended about 3.8 billion years ago. Both bodies also have a regolith that was generated by the continuous rain of particles of all sizes onto the surface. This would also be true of any other bodies that did not have appreciable atmospheres or weathering processes. For example, asteroids and the small moons of Mars also have regoliths. Another similarity between the Moon and Mercury are the large areas of smooth plains associated with impact basins. Although both of these deposits may be lava, the compositions may be significantly different.

7.2.2 Mercury is unique

The differences from the Moon are many and significant. Mercury has the largest iron core (~75% of its radius) compared with its size of any planet or satellite in the Solar System. Although the Moon may have an iron core it is very small compared with other planets and satellites. Mercury is the only terrestrial planet other than Earth that has a dipole magnetic field or a very strong remanent field. The Moon has none at the present time, although it may have small areas of remanent magnetization. Mercury has large areas of intercrater plains in the highlands. In fact, they are the major terrain type on the part of Mercury imaged by *Mariner 10*. Although the Moon does have some patches of intercrater plains they are extremely small (~6% of the surface) compared with those on Mercury (~45%). Mercury displays a widespread, possibly global, distribution of thrust faults that form a unique tectonic framework unlike any other planet or satellite. The basin floor structure of the Caloris basin is unique and has no counterpart elsewhere in the Solar System. The smooth plains and other geologic units have higher albedos than comparable units on the Moon. This may indicate significant differences in composition. Unlike the Moon, Mercury displays strong radar-backscattering patches in the permanently shaded areas of the polar regions. Also, unlike the Moon, there is a large unique radar feature (Feature C) that has no counterpart elsewhere in the Solar System. Finally, the origin of Mercury must be totally different from the Moon. The Moon was probably formed as a result of a large planet-sized impact with the Earth near the end of the final accretion of the planets. This is obviously not the case for Mercury. In fact, the origin of Mercury's enormous iron core requires an origin far different from that of the Moon.

These great contrasts in characteristics between the Moon and Mercury require different processes and/or intensities for shaping the two bodies. Furthermore, their histories must be quite disparate in order to explain these differences. Although comparisons between the Moon and Mercury can be useful, one must use extreme caution in taking these comparisons too far when considering their histories.

7.3 MAPPING MERCURY

7.3.1 The coordinate system

All maps require a coordinate system consisting of latitudes and longitudes by which features are located. The location of the prime meridian (0°) is completely arbitrary.

On Earth, the prime meridian passes through the Greenwich Observatory on the outskirts of London, because it was here that most of the observations required to define longitudes had been made. Only on Earth and the Moon are the longitudes measured 180° east and west of the prime meridian. On all other planets and satellites longitudes are measured 360° east or west of the prime meridian.

The prime meridian on Mercury was selected to pass through the subsolar point when the planet is at perihelion. Because of Mercury's 3:2 resonance between its orbital and rotational period, one hemisphere faces the Sun at one perihelion passage, and the other hemisphere faces the Sun at the next perihelion passage. Consequently, there are two perihelion subsolar points 180° apart. To resolve this ambiguity, the International Astronomical Union in 1970 defined the prime meridian to pass through the subsolar point at the first perihelion after 1 January, 1950.

Mariner 10 did not see the area containing the prime meridian because it was 10° into the night side during the three encounters. However, the center of a small, well-defined crater observed on one of the high-resolution images was calculated to lie within 0.5° of the 20° meridian as defined by the International Astronomical Union's perihelion convention. It was decided that the center of this crater would exactly coincide with the 20° meridian, and to serve as a reference for locating all other longitudes. This 1.5 km diameter crater was called Hun Kal – the number 20 in the ancient Mayan language (Figure 7.2). The Mayans were the most advanced

Figure 7.2. This high resolution *Mariner 10* image shows the position of the small crater Hun Kal (1.5 km diameter), which was used to define the 20° meridian on Mercury.

astronomers in the ancient Americas and used a numbering system based on twenty, rather than the base 10 used in Western civilization. The coordinates of many features were determined from the spacecraft *Ephemeris* (spacecraft position at various times). These were then used to position the latitude–longitude grid with respect to the topography. Longitudes were measured from 0 to 360° increasing to the west. Recent high-resolution radar images of the polar regions have been used to position this latitude–longitude grid much more accurately. These observations show that the pole position is inacurate by 65 ± 2 km from the position shown on the current maps. The new pole position is accurate to 0.05°. The coordinate system on current maps needs to be adjusted by 65 kms.

7.3.2   Naming features on Mercury

The surface of Mercury is divided into 15 areas called quadrangles. Figure 7.3 shows the position of the quadrangles used to identify regions of Mercury. Each map is designated by the letter H (for Hermes, the Greek equivalent of Mercury) followed by a number from 1 to 15. Nine of these quadrangles viewed by *Mariner 10* have been compiled into shaded relief maps at a scale of 1:5 million. Unfortunately large parts of three of these quadrangles were not imaged by *Mariner 10* so they are very incomplete. The nine quadrangles are further designated by the names of prominent surface features contained in the areas. For example, the south polar map is called the Bach (H-15) quadrangle since there is a large crater in the area by that name.

Craters on Mercury are named after famous people in the arts, including artists, authors, and musicians, such as Dickens, Michelangelo, and Beethoven. There are, however, exceptions. The crater Hun Kal is named after the number 20 in the Mayan language as explained earlier. Also the bright rayed crater Kuiper is named after a famous astronomer who was a member of the *Mariner 10* science team before his untimely death in December 1973. Prominent ridges or scarps (called "rupes" from Latin) are usually named after ships of exploration and scientific research, such as *Discovery* and *Victoria*. Exceptions are two prominent ridges named Antoniadi and Schiaparelli for the astronomers who first mapped Mercury from Earth-based observations. Valleys (called "valles" from Latin) are named after prominent radio observatories such as Arecibo and Goldstone. Plains (called "planitiae" from Latin) are named after the word for the planet Mercury in various languages, and for gods from ancient cultures who had a role similar to that of the Roman god Mercury. Typical names are Odin (Scandinavian) and Tir (Germanic). Borealis Planitia (Northern Plains) and Caloris Planitia (Plains of Heat) are exceptions.

7.3.3   Maps and topographic representations

There have been a number of maps prepared from the imaging data. The map shown in Figure 7.4 is a shaded relief map of Mercury's Shakespeare quadrangle. This type of map is prepared by highly trained artists using airbrushes.

An *Atlas of Mercury* was compiled from *Mariner 10* images and is the most comprehensive set of maps of the *Mariner 10* coverage. The atlas consists of

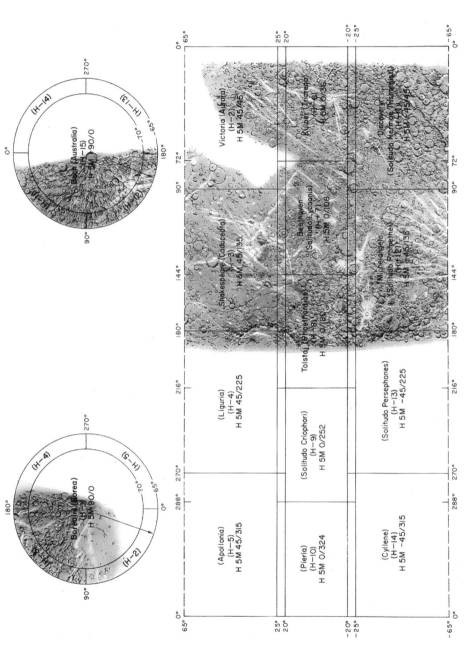

Figure 7.3. This map shows the distribution and names of the 15 (1:5 million scale) quadrangles (also on the CD). Only nine of these quadrangles were wholly or partly imaged by *Mariner 10*. About 55% of the planet remains unexplored (from Atlas of Mercury).

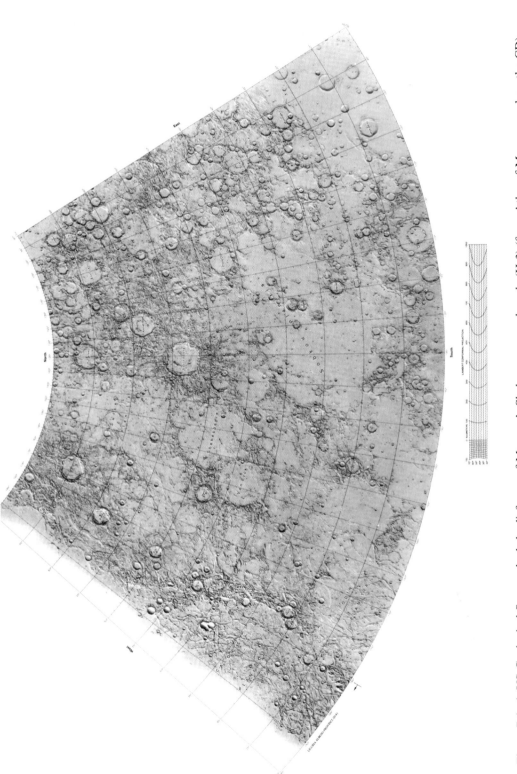

Figure 7.4. A US Geological Survey shaded relief map of Mercury's Shakespeare quadrangle (H-3) (from Atlas of Mercury and on the CD).

Figure 7.5. Photomosaic of Mercury's south polar region, taken by *Mariner 10* on its second encounter.

photomosaics and shaded relief maps of each quadrangle plus individual images of interesting features. It also includes some stereo images acquired from the first and second encounters. Figure 7.5 shows one of the photomosaics from this atlas, and Figure 7.6 is a photomosaic of the entire equatorial region. The accompanying CD contains all of the photomosaics, airbrush maps, and the best single images from *Mariner 10*.

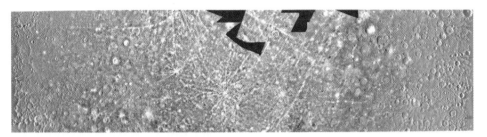

Figure 7.6. Photomosaic of the equatorial region of Mercury from about 10 to 190° longitude and ±30° latitude; the only mosaic from terminator to terminator. The photomosaic is a Mercator projection (courtesy of Jet Propulsion Laboratory).

In addition to the maps mentioned above, there are also special geologic maps of nine quadrangles. These maps are an interpretation of the geology of Mercury viewed by *Mariner 10*.

7.4   RADAR CHARACTERISTICS AND SPECIAL FEATURES

Ground-based radar observations from Arecibo, Puerto Rico, the Very Large Array (VLA) in Soccoro, New Mexico, and the Goldstone radar facility in California's Mohave Desert have contributed significantly to our knowledge of Mercury's surface. As usual, observations of Mercury are not easy. The radar disk of Mercury is comparable to a dime at 16,000 km distance! At the same time the data have been so interesting, some with multiple interpretations, that new and exciting questions have been raised.

7.4.1   Roughness, and equatorial and low latitude topography

Important radar observations of Mercury were made at the Goldstone radar facility at Fort Irwin, California in the 1970s and 1980s. At Mercury's closest approach to Earth, features down to 10 km in size could be resolved at 12.5 cm (S-band) and 3.5 cm (X-band) wavelengths. From these observations it was possible to obtain topographic profiles along equatorial regions on both the side imaged by *Mariner 10* and the unimaged side. For the terrains measured by these observations, it was found that crater rim heights were generally low and crater floor depths relatively shallow. Total topographical height differences greater than 2 km are uncommon at the spatial resolution of the radar.

Radar observations from Arecibo Observatory also measured *altimetric profiles* and identified large-scale topographic features such as subsidence zones, highlands, and smooth plains. A large, 2.5 km deep crater at 279°, 8.5°N is one of the largest structures discovered on the unimaged side.

Surface roughness and *transparency* may also be determined from radar observations. Mercury and the Moon both have average *radar cross-sections* of about 0.06 at the 13 cm wavelength. For comparison, Venus has a cross section of about 0.11, and Mars a range from about 0.04–0.15 at the 13 cm wavelength. Solid rock surfaces have cross sections of 0.15–0.25 and rock powders about 0.03–0.06. Thus, most of Mercury's surface is dominated by relatively porous regolith, rather than solid rock. This is consistent with measurements of Mercury's microwave and infrared emissions, and its photometry. Microwave observations from 0.3 to 20.5 cm indicate that Mercury's regolith is at least two to three times more transparent than the lunar maria and at least 40% more transparent than the lunar highlands. This difference is likely due to a lower abundance of iron and titanium in Mercury's regolith (see Chapter 8). Mercury's quasispecular roughness measurements (mean slopes) are very similar to the Moon's which show differences between the smooth mare surfaces (4°) and the lunar highlands (8°). These same differences are seen for Mercury's smooth plains (5.3°) and the rougher intercrater plains (8.3°). Like the Moon, there is no evidence for the extremely smooth terrain (~1°) that has been seen in some areas of Mars.

7.4.2   The Goldstein features

In 1970, two very large radar features were discovered on Mercury by Richard Goldstein of the Jet Propulsion Laboratory using the Goldstone radar facility. Circularly polarized, monochromatic waves were beamed at Mercury. Then the sense of circular polarization was reversed and more waves reflected from Mercury. The reflection from the "opposite sense" circular polarized beams revealed large radar-bright features. Two definite features were identified, with the possibility of a third. Because of north/south and spreading ambiguities inherent in the radar imaging at that time, identification of the exact location and the nature of the three radar-bright topographic features would have to wait for two decades.

   The first full-disk radar imaging of Mercury was done using a direct interferometric imaging approach which does not have the north/south ambiguity of the earlier methods. The Goldstone radar facility was used to transmit the maximum signal possible to Mercury and the VLA receivers near Socorro, New Mexico were configured to receive it. By spacing the array of receivers in an optimal configuration, it was possible to obtain high spatial imaging of the surface radar reflectance at a resolution of 15 km.

   One of the Goldstein features was identified with a large equatorial feature near 240°W longitude. The other Goldstein feature was resolved into two separate mid-latitude features, one in the north and one in the south. The extent of the uncertainty in the Goldstone observations became understood when it was discovered that the northern and southern features in the Goldstone/VLA image had the same longitude, close to 345°W. The three topographic features have been given the designation of radar-bright regions A (345°W longitude, −32° latitude), B (345°W longitude, 58° longitude), and C (240°W longitude, 0° latitude).

   Very high-resolution (1.5–3 km) radar images at 12.6 cm wavelength obtained by the Arecibo radar facility show that both features A and B are relatively fresh impact craters with radar-bright ejecta blankets and rays. On the Moon only Tycho and Copernicus show radar-bright rays at 3.8 cm wavelength; at 70 cm wavelength they do not show at all. The radar-bright rays are probably rough ejecta and small fresh secondary impact craters that are rough at the radar wavelength of the observations. In fact, the fresh rayed craters Kuiper and Copley, seen on *Mariner 10* images, are also radar-bright. Features A and B are seen in Figure 7.7. Feature A is about 85 km diameter with an extensive ray system and a rough radar-bright floor, consistent with fresh impact crater morphology similar to Tycho or Copernicus on the Moon. Feature B is about the same size (~87 km) with radar-bright rays and a radar-dark floor. The radar-dark floor indicates it is smooth at 12.6 cm wavelength, contrary to Feature A. Feature C consists of a large circular region about 1000 km diameter consisting of small radar-bright spots. There appears to be no central structure as in Features A and B. The geologic nature of this feature is not known, but recent radar images suggest it may be a swarm of impact craters or possibly secondary impacts from an obscure crater.

7.4.3   Radar observations discover highly backscattering polar deposits

Not only did the radar imaging confirm the Goldstein features but it discovered a north polar "anomaly". There was a small area in the polar regions where a very high fraction of the incident radar was reflected back to Earth with the same characteristics as reflections from Mars' south polar cap and from the icy Galilean satellites. The signal intensity and the location suggested that Mercury had polar water ice. Follow-up observations were made with the Arecibo radar facility in Puerto Rico. While a different type of imaging was used, the results were equally interesting; not only was the north polar "anomaly" confirmed but a new south polar "anomaly" was discovered. The radar bright regions can be seen in Figure 7.7, and a map of the polar deposits is shown in Figure 7.8.

   If these deposits are indeed ice, then the Moon and all terrestrial planets, except Venus, have ice deposits in their polar regions. The Greenland and Antarctic ice caps on Earth are the remnants of the last ice age that ended about 12,000 years ago. The source of the ice is, of course, Earth's oceans. The polar caps of Mars consist of both water (H_2O) and carbon dioxide (CO_2) ices. During Martian summers at the poles the CO_2 sublimes away leaving a residual cap of water ice. Recent spacecraft measurements from the *Odyssey* mission's neutron and gamma-ray spectrometer have discovered very large amounts of buried water ice down to about 60° latitude in the southern hemisphere, and in some places down to ~40° latitude in the northern hemisphere. The source of this water ice may also have been past oceans on Mars. The neutron and gamma-ray spectrometers on the *Lunar Prospector* spacecraft discovered enhanced hydrogen (H) signals in permanently shadowed craters in the polar regions of the Moon. This has been interpreted as water ice with a concentration of $1.5 \pm 0.8\%$ weight fraction.

   The discovery of possible water ices at high latitudes (72°–90°N and S) in the polar regions of Mercury astounded planetary scientists and the public alike. How could this hot planet, with daylight lasting 88 Earth days and temperatures high enough to melt lead possibly have water ice at the southern and northern polar areas? More observations were taken at the same two facilities, and an upgrade

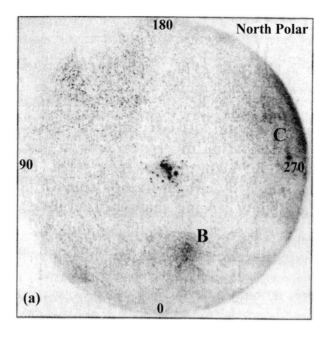

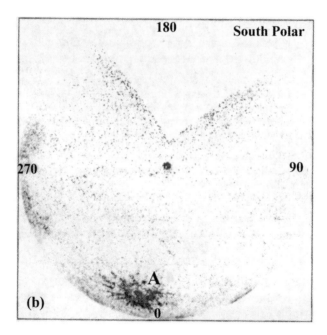

Figure 7.7. Arecibo *depolarized* radar images of the (a) northern, and (b) southern hemispheres of Mercury, in polar projection. The radar features A, B, and C are indicated on the figure. The polar deposits can also be seen near the poles (from Harmon, 1997).

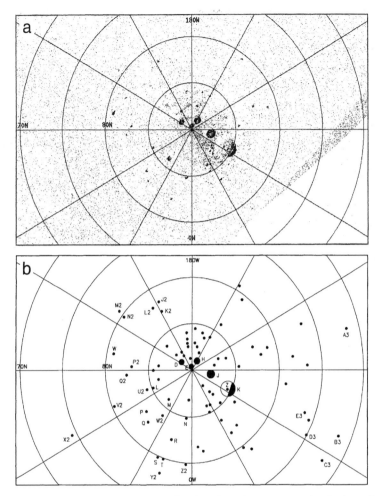

Figure 7.8. Bright radar signals from localized regions in permanently shadowed craters at high northern latitudes on Mercury's surface are shown. The actual radar image is shown in (a) with a superimposed location grid. A detailed map and numbering scheme are shown in (b) (from Harmon *et al.*, 2001).

on the Arecibo radar facility permitted even higher spatial resolution observations. Soon maps had been made of the north and south polar and high-latitude regions, that showed many locations of high radar backscatter coming from what was determined to be permanently shadowed regions in the interiors of craters. The reason Mercury has permanently shadowed craters is that its obliquity is essentially zero. Therefore, the rotation axis is perpendicular to its orbital plane around the Sun, so there are regions in craters at high-latitudes that never see the Sun. The deposits are concentrated only in the freshest craters, and even in some craters less than 10 km in diameter. Degraded craters do not show the highly radar backscattering deposits, probably because there is no permanently shadowed regions in these

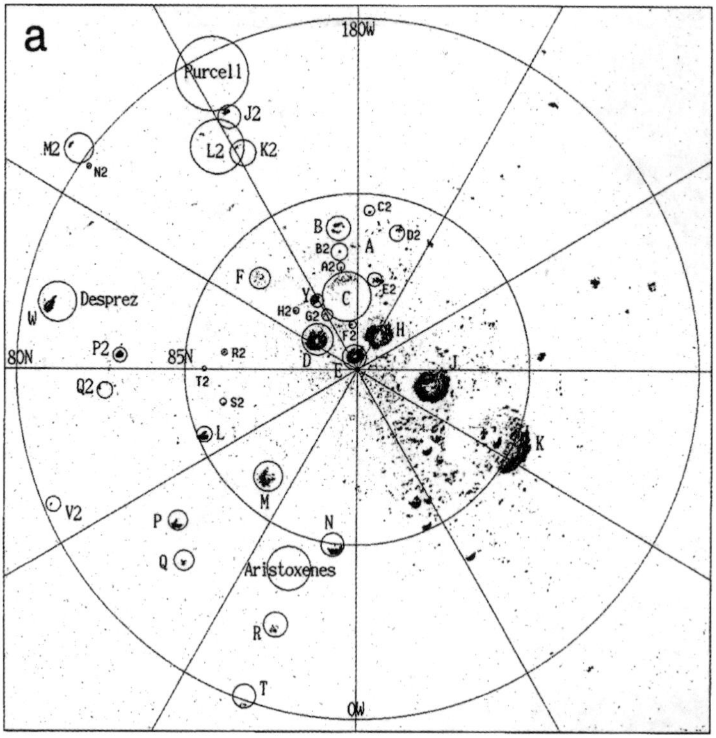

Figure 7.9. Detailed high-resolution radar image showing the radar-bright deposits in the north polar regions, and with a superposed map of their locations and designations (from Harmon *et al.*, 2001).

low-rimmed and shallow craters. In fact, the permanently shadowed cold traps are essentially full. Furthermore, the strong radar signal indicates that the material is relatively pure. The estimated thickness of the deposits is believed to be approximately between 2 and 20 m. The upper limit is, in fact, arbitrary because the radar data cannot put a upper limit on the thickness. The area covered by these deposits (both north and south) is estimated to be $\sim 3 \pm 1 \times 10^{14}$ cm^2. This would be equivalent to 4×10^{16} to 8×10^{17} g of ice, or 40–800 km^3 for a 2–20 m thick deposit. Each meter thickness of ice would be equivalent to about 10^{13} kilograms of ice.

Not only is the observational evidence strong for water ice but so are the theoretical models and calculations that have been made following the discovery. It has been calculated that water ice can be stable in the interiors of craters even down to 72° latitude if covered with only a few centimeters of dusty regolith, or if it is relatively new. This means that water ice may still linger in its perpetually shadowed craters or even in illuminated craters at high latitude if covered with a veneer of regolith. Figure 7.9 shows the actual radar image with the highly backscattering radar signals for the northern hemisphere. A map of those locations with a

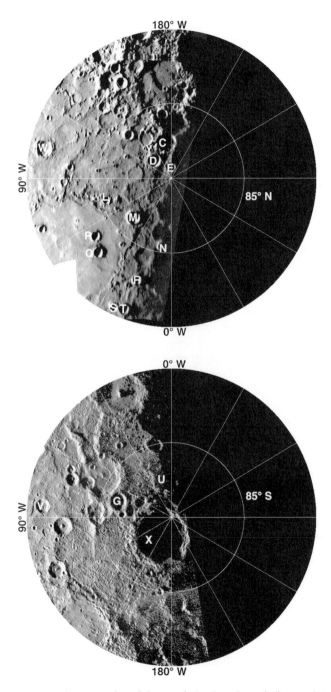

Figure 7.10. *Mariner 10* photomosaics of the north (top) and south (bottom) polar regions of Mercury showing some of the larger craters with radar bright deposits (lettered as on Figure 7.9) (courtesy Mark Robinson, Northwestern University).

numbering identification scheme is overlain on the image. Figure 7.10 includes *Mariner 10* photomosics of the north and south polar regions showing some of the larger craters with radar bright deposits.

The current evidence suggests that Mercury's polar deposits are probably water ice. The most likely sources of the water are micrometeorite, comet, and water-rich asteroid impacts. Extrapolating the current terrestrial influx of interplanetary dust particles to that at Mercury indicates that continual micrometeorite bombardment of Mercury over the last 3.5 billion years could have delivered $3-60 \times 10^{13}$ kg of water ice to the permanantly shadowed polar regions (an average thickness of 0.8–20 m). Impacts from Jupiter-family comets over the last 3.5 billion years can supply $0.1-200 \times 10^{13}$ kg of water to Mercury's polar regions (corresponding to an ice layer between 0.05–60 m thick). Halley-type comets can supply $0.2-20 \times 10^{16}$ g of water to the poles (0.1–8 m ice thickness). These sources provide more than enough water to account for the estimated volume of ice at the poles. The ice deposits could, at least in part, be relatively recent deposits, if the two radar features A and B were the result of recent cometary or water-rich asteroid impacts.

While the evidence for water ice is strong, other possibilities for the material causing the high radar backscatter signal have been suggested. Unfortunately, until this discovery there has not been much need for ground-based laboratory experiments to determine the radar properties of planetary materials. During and following World War II, there were some measurements of a variety of materials of interest to the military. Among them was sulfur ($S_n, n = 2, 4, \ldots$), a substance used as an electrical insulator. One property of a good radar backscattering material is that it is a good electrical insulator, sulfur is such a substance. A source of sulfur is the constant rain of meteoritic material. The problem with sulfur being the deposits on Mercury is that it is stable at higher temperatures than water, and there are no highly radar backscatter deposits in the polar regions where temperatures are within the stability range of sulfur. A 1-meter thick layer of water ice is stable for one billion years at a temperature of $-161°C$ while sulfur is stable at a considerably higher temperature of $-55°C$. Much of the region surrounding permanently shadowed craters is less than $-55°C$, but there are no radar reflective deposits there. Very cold silicate glass has also been suggested as a possibility. The *MESSENGER* mission to Mercury should be able to address this problem from critical measurements made during its orbital lifetime.

8

Surface composition

8.1 ALBEDO AND COLOR

Mariner 10 made no measurements that could determine the elemental abundances, specific minerals, or rock types on Mercury. All we know about Mercury's surface composition comes from ground-based observations and inferences from color reconstructions of *Mariner 10* images. Recalibrations of the *Mariner 10* images and a technique of ratioing images and looking at relative brightness from two different colors has resulted in several important new insights into the makeup of the *regolith* and possibly its iron oxide (FeO) content. They have also suggested the location of compositional boundaries.

Albedo is one word used to quantify the percentage of light reflected back from a surface. The albedo of Mercury's surface varies from one location to another and from one wavelength to another. The human eye is sensitive to the spectral range from about 400 to 700 nm and perceives what we call the *visible spectrum.* The colors violet, indigo, blue, green, yellow, orange, and red all fall within that range. Other wavelength ranges are also referred to as having colors, just not visible colors. One way planetary surfaces are compared in their scattering and compositional characteristics is by their color. Often the color may refer to the relative albedo at one spectral region to the albedo of a different planetary surface (or atmosphere) at the same wavelength interval. The albedo of a surface will also vary when the angle of incidence and exitance of reflecting sunlight changes. Thus, to properly compare different albedo measurements, not only must the location be known but also the illumination geometry must be the same on both surfaces being compared. Thus, researchers must be careful to make comparisons that are justified within the experimental uncertainties.

8.1.1 Mercury's EUV, UV–VIS, and near-IR albedo

Observations of the day side albedos of the Moon and Mercury were made by *Mariner 10* with the airglow spectrometer (extreme ultraviolet–[EUV]) and the imaging experiment (3300 Å to ultraviolet [UV]). Analysis of these data showed Mercury's surface to have lower overall EUV albedo than the Moon by a few percent. In addition, Mercury's EUV albedo is somewhat higher at the shorter wavelengths (600–1000 Å) than at longer wavelengths (1000–1650 Å). However, the albedo rapidly rises in the UV from 1600–4000 Å. It continues to rise through the visible and the near-infrared (IR) to 10,000 Å. Different regions on the surface have different albedos at the same wavelength. For example, fresh craters and crater rays are very bright, having higher albedos than the darker surrounding regions.

The composition of the surface materials plays an important role in explaining the albedo of the surface. So does the grain size and porosity of the surface materials. For example, finely crystalline silicates low in iron and titanium tend to be brighter and scatter more light off the surface.

Mariner 10 obtained many images of the surface with the visible light filter (V) that is centered on 554 nm. There were enough images obtained to permit comparison of albedos from many locations. These measurements were compared to albedo measurements of the Moon made at 5° phase angle by using modeling techniques. Table 8.1 compares the albedo from regions on Mercury's surface to similar landforms on the Moon.

From Table 8.1, it appears that Mercury has a higher albedo in the visible (VIS) wavelength range than the Moon at some locations. This may indicate important compositional differences between Mercury and the Moon. In addition, Mercury albedos tend to be more uniform in the UV–VIS than those of the Moon which vary greatly from maria to highlands. Exceptions to the uniform albedos are the brightest rays on Mercury which show the highest albedos in Table 8.1.

Table 8.1. Comparison of albedo from regions on Mercury's surface to similar landforms on the Moon.

Terrain	Albedo at 554 nm wavelength (visable) normalized to 5° phase angle
Mercury Caloris smooth plains	0.12–0.13
Lunar maria	0.06–0.07
Mercury highlands (intercrater plains)	0.16–0.18
Lunar highlands	0.10–0.11
Mercury bright rayed craters	0.36–0.41
Lunar bright rayed craters	0.15–0.16

8.1.2   Changes of albedo with phase change

It has been known for centuries that as the Moon changes in phase from a thin crescent to full, the brightness of the surface increases dramatically. This increase in brightness is much more than could be accounted for by just having more of the surface illuminated by the Sun as the *phase angle* decreased (the Moon approached full phase). The extra light coming from the surface is, in part, a consequence of a decrease in the amount of shadowing of reflected light from the surface as the light path becomes more and more direct (perpendicular to the surface) from the Sun. A decade ago, Bruce Hapke, a planetary scientist at the University of Pittsburgh, showed that the increase in light near full phase is also because the light entering the soil particles scattered out toward the observer with the same electromagnetic wave properties as it entered. This phenomenon, called coherence, increases the amplitude of the light and contributes to the surge of light near full phase. The overall shape of the phase curve is diagnostic of backscattering efficiency, particle size, microscopic roughness, and other properties of the surface materials.

During the past few years, Mercury was observed by instrumentation on the *SOlar Heliospheric Observer* (*SOHO*) from Earth orbit, obtaining several new data points near 180° phase angle (thin crescent). These added points have enabled scientists to study the entire phase behavior of both the Moon and Mercury. Mercury exhibits slightly more backscattering than the Moon near full phase (has a sharper peak near 0° phase angle). This can be seen in Figure. 8.1. The curves in the figure are based on models of the scattering behavior of the powdered surface of an airless planet. Such models make predictions on the physical and optical structure of the small grains in the topmost 1 mm of the surface, which is the region where sunlight is scattered, reflected, and absorbed by the particles before eventually reaching the telescope. From such modelling, we have reason to believe that there are other differences as well between the regolith of the Moon and Mercury. For example, it seems that Mercury's surface grains are smaller and more transparent than their lunar counterparts. They also seem to reflect light more effectively towards the direction of the Sun, which may be due to the presence of complex or fractured grains. Particles with these properties are known from the Moon as glassy conglomerates of rocky grains called *agglutinates*, formed during heat generating impacts. The inference is that such particles are more common on Mercury, are less dark, or are more complex. The high *translucency* goes hand in hand with results from other studies that indicate that the surface is generally much poorer in iron and its compounds, darkening constituents, than the Moon's surface. In addition, Mercury's surface seems to be smoother than the Moon's.

8.1.3   Spectral slope and "maturity"

Reflected light from silicate regoliths of airless bodies like Mercury and the Moon have been studied for decades with the hope of isolating specific physical properties

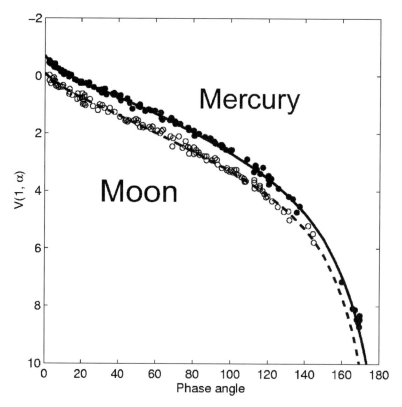

Figure 8.1. Phase curves for Mercury and the Moon. The total light coming from the planet as it changes phase from crescent to full is measured and plotted. The brightness of Mercury (top) and the Moon (bottom) in the V filter band (visible wavelengths) is scaled to the same distance (1 AU) to the Sun and the Earth. Though the two best-fit curves drawn through the observed data points are similar they are not exactly the same, which indicates different light scattering behavior for the two bodies. There is uncertainty in the absolute brightness when the two curves are adjusted for the actual size of the Earth-facing disk. Such an adjustment has not been made for the curves shown here (courtesy of Johan Warell, University of Uppsala, Sweden).

that can be inferred from spectral features and from *spectral slope*. The task has not been an easy one because there are many variables such as composition and grain size that vary among the terrestrial planets. In addition, the amount of glassy material formed by impact, the effects of impacting ions and electrons on color and crystalline lattice structure, the degree to which the surface is protected from charged particles by magnetic fields, and the age of the regolith also confound an easy answer. But, because of the tenacity of researchers seeking to understand distant regoliths in the Solar System, much progress has been made. Today most researchers would agree that spectral slope is an indicator of the degree to which the surface has been altered by external processes such as the

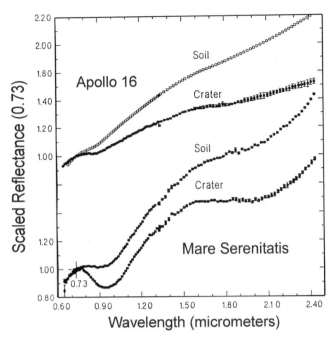

Figure 8.2. Regions on the Moon that have similar composition have different spectral slopes because of different amounts of space weathering. The slope is steeper from shorter to longer wavelength (left to right) as the surface becomes more mature (adapted from Pieters and Englert, 1993).

ones listed above. Such alterations are called space weathering (for details refer to the discussion at the end of Chapter 5.) Freshly exposed rock and soil from a recent impact has not experienced much space weathering, while ancient terrains are highly altered by space weathering. We call the degree to which the surface has been altered, the maturity of the surface. A soil, freshly exposed by an impact is an *immature* soil. A soil that has been exposed to the effects of space weathering for two billion years is said to be *mature*.

Spectral slope is a major indicator of soil maturity. Most of our knowledge regarding this relationship between slope and maturity comes from studying the Moon and to a lesser degree, the asteroids. The spectral slope of many different lithologies on the Moon is well known. Spectra of small areas on the Moon with similar compositions have different slopes because of different maturities. This is illustrated in Figure 8.2 where lunar soils in and near to craters are compared to soils undisturbed by the impact.

Mercury's *reflectance* between 400 and 1000 nm has been very well measured in the last decade because such measurements can be made from ground-based telescopes with modest size mirrors and grating spectrographs. Recent photometric observations of the Earth-facing hemisphere of Mercury not imaged by Mariner 10 were obtained in 1997 and 1998 with the Swedish Vacuum Solar Telescope on

La Palma in the Canary Islands. The spectrum of Mercury was found to have a linear slope from 650 to 940 nm, indicating that the average Mercurian regolith is considerably more mature than "near-side" lunar immature anorthositic regolith. These observations also found that Mercury's surface is more backscattering than that of the Moon at these wavelengths. This wavelength range is discussed in more detail in the sections below.

8.2   MATERIALS OF TERRESTRIAL PLANETARY SURFACES

The most common solid materials on Earth are rocks and soils that are made up mostly of oxygen (O) and silicon (Si). Many minerals are *solid solutions*, in varying proportions, of the oxides of silicon (Si), calcium (Ca), aluminum (Al), magnesium (Mg), sodium (Na) and potassium (K). Rocks formed from such minerals are called *silicates*. Minerals form with specific combinations of these elements based upon atomic radius, distribution of electrons, and other important characteristics of chemical bonding. Iron plays an especially interesting role in that it can be present as a component of silicates (oxides) or it can be present as a separate metallic phase along with nickel, sulfur and certain other elements. Common rock-forming minerals include feldspar, pyroxene, olivine, and iron-titanium oxides.

On Earth, feldspars are the most abundant of all minerals in the crust. While closely related, they fall into groups: potassium and barium (Ba) feldspars, or sodium and calcium feldspars for example. The types have different crystalline structures. The structure of feldspars is a continuous three-dimensional network of SiO_4 and AlO_4 tetrahedra. The structure is elastic and adjusts itself to the size of cations that occupy the interstitial openings of the chain. If the cations are large (K, Ba), the feldspar formed may be orthoclase ($KalSi_3O_8$). If the cations in the interstitial openings are small (Ca, Na), the feldspar may be albite ($NaAlSi_3O_8$), or Ca-rich anorthite ($CaAl_2Si_2O_8$). Orthoclase, albite, and anorthite make up a three component system – Or, Ab, and An are the shorthand designation of the three components. The way in which the two or three components can mix and form solid solutions with different proportions of each is complicated. It depends upon temperature, pressure, composition of the mixture, and the relative abundance of K, Na, Ca, Al, and other elements. The description of the solid solution is written with the shorthand designation of the components with a subscript denoting the percentage of the component. For example, $Or_{26}Ab_{66}An_8$, is the designation of a feldspar (called anorthoclase) formed at high temperature.

One common solid solution series is plagioclase feldspar $(Ca,Na)(Al,Si)AlSi_2O_8$. Plagioclase feldspar is formed both at high and low temperatures but it contains only two components of the three component feldspar system, Ab and An. Na-rich plagioclase feldspar would have a designation such as $Ab_{95}An_5$ and be composed of 95% $NaAlSi_3O_8$ and 5% $CaAl_2Si_2O_8$. Such a feldspar is called albite. Another example is Ab_4An_{96} – 4% $NaAlSi_3O_8$ and 96% $CaAl_2Si_2O_8$. Such a feldspar is called anorthite. Any other combination is also possible. Plagioclase feldspars made up of different solid solutions are given specific names which indicate the

Table 8.2. Names of the divisions of the plagioclase series showing the abundance, range for the Ab end member component.

Plagioclase feldspar type	Ab end-member (%)
Albite	Ab_{100} to Ab_{90}
Oligoclase	Ab_{90} to Ab_{70}
Andesine	Ab_{70} to Ab_{50}
Labradorite	Ab_{50} to Ab_{30}
Bytownite	Ab_{30} to Ab_{10}
Anorthite	Ab_{10} to Ab_{0}

relative percentages of each component. Table 8.2 gives the names of the divisions of the plagioclase series showing the abundance range for the $NaAlSi_3O_8$, Ab *end member component*.

The pyroxene group is also an abundant group of minerals on Earth. The members of the group are closely related to one another, like the members of the feldspar group, and also crystallize in two different systems: orthorhombic and monoclinic. Again, the possibilities of different types and compositions are many and the cause of the differences are related to temperature, pressure, and abundance of elements present in the mix at the time of formation. Basically, there are Fe-rich, Mg-rich, Ca-rich, NaAl-rich, CaMn-rich, and "other"-rich pyroxenes. Examples of ortho-pyroxene are enstatite ($MgSiO_3$) and hypersthene ($(Mg,Fe)SiO_3$). Examples of clino-pyroxene are diopside ($CaMgSi_2O_6$), hedenbergite ($CaFeSi_2O_6$), and augite (intermediate between diopside and hedenbergite with some Al). There are others but this serves to give a general idea.

Another mineral series that is very common on Earth, especially in the mantle, is olivine ($(Mg,Fe)_2SiO_4$). Like the feldspar system, it is a continuous solid solution of two end member components, Mg_2SiO_4 and Fe_2SiO_4. They are called forsterite (Fo) and fayalite (Fa), respectively. The shorthand notation is similar to that for the plagioclase solid solution series. For example, an olivine composition might be $Fo_{66}Fa_{34}$.

Other minerals are sulfates, sulfides, carbonates, amphiboles, and micas, to name only a few. Earth has a rich diversity of thousands of minerals owing to its water-rich environment, plate tectonics, long volcanic history, and hydrologic cycle. Iron-titanium oxides are plentiful on the Moon but to date there is no evidence for their existence on Mercury. Mars has considerable water ice in its sub-surface material and at the polar regions. There is evidence for past flowing water and possible oceans on its surface, and surface weathering of its minerals, rocks, and soils. Mars soils are almost surely more diverse than the regoliths of the Moon and Mercury. They are certainly more altered by water processes than the regoliths of Mercury or the Moon. Mercury and the Moon, with no sign of plate tectonics or a hydrologic cycle, has an assemblage of rocks, minerals, and regoliths, probably limited to those that are associated with crystallization of *magma, extrusive* lava flows, possible *igneous* intrusions, and meteoritic impact melting, fracturing, and mixing.

8.3 MERCURY'S SURFACE COMPOSITION

The past two decades of Mercury surface observations have resulted in many good observations that are continuously being refined and augmented. These observations include imaging in visible light (400–700 nm), near-IR (700–1000 nm) and mid-IR (2.5–13.5 nm) spectroscopy. Certain wavelength regions are blocked, or at least attenuated because of Earth's atmosphere. Terrestrial atmospheric NO_2, O_3, CO_2, CO, H_2O, and CH_4, are the main absorbers between 400–1000 nm and 1.0–13.5 nm. Regions in the spectrum where light is attenuated by absorption from terrestrial molecules are called *telluric absorptions*. Some remove more of Mercury's light on its path through the atmosphere than others and in many cases proper corrections for the absorptions can be made to the spectra from Mercury so that they can be interpreted properly in terms of rock and mineral signatures. Sections of the Earth's atmosphere where no, or very little light can penetrate are called *opaque* to radiation and these spectral regions result in gaps in our knowledge about Mercury's spectrum. Regions of the spectrum where little or practically no attenuation of light by the Earth's atmosphere occurs are called *atmospheric windows*.

We have only a little knowledge of the chemical makeup of Mercury's surface. Thus, we shall limit our discussion to only the most common of rock types and rock forming minerals. A good candidate for a rock type on Mercury is low-iron basalt (see Chapter 10). The dominant rock type on the surfaces of Earth, Venus, and Mars is basalt. Basalt is a rock primarily composed of feldspar, pyroxene, and olivine. Another possible rock type on Mercury is anorthosite, a rock composed of more than 90% plagioclase feldspar with other component minerals like pyroxene. These are the two main rock types on the Moon. There is a wide range of possible compositions in both basalts and anorthosites.

Basalts and anorthosites have very different origins. Basalt is formed by a *petrogenetic* process called *partial melting*. As the temperature of a parent rock is increased the low-melting point fraction melts first and has a characteristic composition which rises and is erupted on the surface. This composition corresponds to the low-melting point composition of the parent rock. On the Moon both the mare and KREEP (potassium (K), rare earth elements, phosphorous (P)) basalts have characteristic low-melting point compositions that were produced by partial melting of a less differentiated rock in the interior of the Moon. Anorthosite, on the other hand, is formed by a process called *fractional crystallization*. In this process, when an igneous melt begins to cool and crystallize, the denser crystals sink through the residual liquid and the lighter (less dense) crystals float to the top. Therefore, layers accumulate in which one mineral is greatly concentrated. The lunar highlands anorthositic rocks are characterized by a super-abundance of one mineral, anorthite ($CaAl_2Si_2O_8$). This composition is far removed from low-melting point liquids and was formed by fractional crystallization.

Recall, from the discussions of albedo and slope, that Mercury's surface composition is different from that of the Moon even though they resemble one another morphologically. The compositional differences are likely to be rather subtle, however, like the relative amounts of Na, or K in the feldspar and the relative amounts of Mg, or Fe in the pyroxenes or their crystalline structure. But it is likely that Mercury has igneous rocks that either crystallized in place or flowed from fissures early in its history.

8.3.1   Measuring surface composition with a telescope

Planetary astronomers use instrumentation on telescopes for compositional analyses of planetary surfaces. Light from the planet's surface carries information about the composition and structure of the dust, soils, rocks, and minerals in the regolith. The exact way in which light interacts with the regolith depends upon many factors. Some factors of major importance are: wavelength; temperature and thermal gradients; composition; and grain size. Measurements made in the laboratory of typical soils, rocks, minerals, meteorites, and lunar samples permit an understanding of what spectral features may be predicted for a given composition and grain size.

8.3.2   Visible and near-IR spectroscopy

The power of visible and near-IR telescopic spectroscopy (400–2500 nm) has been demonstrated throughout the solar system for decades. For example, the size and shape of absorption bands in reflected light from the surface caused by crystal field transitions, metal-metal intervalence charge transfer transitions, and oxygen-metal charge transfer transitions of the materials in the regolith, are very diagnostic.

   One of the best known features in reflectance spectroscopy of terrestrial bodies is an absorption caused by an electronic transition in an iron cation (Fe^{2+}). The electronic transition occurs when Fe is bound to O in a lattice of a silicate. Planetary scientists familiar with this absorption band commonly refer to it as the FeO band. It occurs between 900 and 1000 nm in the reflected spectrum from a surface with FeO in the silicate portion of the regolith.

8.3.3   The FeO band in lunar spectra

The origin of the Moon was poorly understood until the late 1980s. In attempting to put together a coherent hypothesis for the lunar origin, it was necessary to know the composition not only of the lunar surface but also of the deeper, more primitive materials of the lunar *mantle*. The lunar mantle is the region below the surface and crust on the Moon that exists undisturbed except by volcanism and deep, excavating impacts.

   Prior to the *Apollo* missions that returned lunar surface samples for laboratory analysis, ground-based telescopic visible and near-IR spectroscopy was the best method for compositional studies. Several absorptions in the reflected light from the lunar surface can be interpreted in terms of the cations in the minerals of the lunar regolith that absorb sunlight. These absorptions by Fe^{2+}, Mg^{2+}, and other cations look different with small changes in the crystalline lattices that make up the rocks and rock debris. The best known of these absorption bands are probably the lunar olivine and pyroxene bands. Mapping the composition of the lunar near-side resulted in the identification of mantle material in Copernicus crater by measuring the depth, width, and the location of absorption minima in the bands. Comparisons with spectra of terrestrial laboratory samples aided in this accomplishment. The absorption features in the spectra shown in Figure 8.2 are examples.

8.3.4   The FeO band in Mercurian spectra

Only a dozen or so locations have been measured but from this evidence we deduce that Mercury's surface materials may contain at most, a few percent FeO in surface materials, and maybe in some locations, none at all. Some near-IR spectra are shown in Figure 8.3 along with some spectra of the Moon's surface for comparison. The

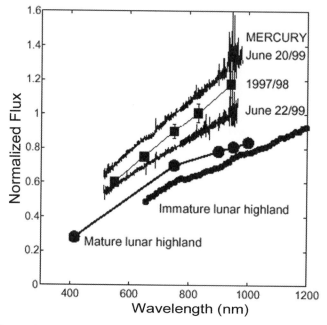

Figure 8.3. Comparison of the reflectance spectra of Mercury's surface with that of the Moon. The slopes of the spectra are similar from 400–800 nm, then the rate of increase of the lunar spectra drops. The Mercury spectra continue to rise at the same rate to 1000 nm. The depression beyond 800 nm in the lunar spectra is due to the presence of FeO in the lunar soil. The fact that there is no such dip in the Mercury spectra is evidence for low or no FeO in the Mercurian regolith at the locations measured and also for a more mature surface than the Moon (courtesy Johan Warell, Mercury spectra from the Nordic Optical Telescope and the Swedish Vacuum Solar Telescope).

feature in the spectrum that indicates the presence of FeO appears as a region of less steep rise in the lunar spectra. If there were measurable FeO in Mercury's soil the absorption would be at the same location on the x-axis, but no dip in the spectrum is seen. What is shown in the steeper slope of the spectrum indicating that Mercury's surface is very mature, more mature than the Moon's. It is very altered by space weathering mechanisms like meteorite bombardment, solar wind implantation, ion sputtering, photon sputtering, and cosmic ray impact and, to some extent, resembles the spectrum of Apollo 16 soil in Figure 8.2.

8.3.5 Mid-IR spectroscopy

Spectroscopic measurements of Mercury's surface at longer wavelengths (2.5–13.5 μm) have provided important chemical information about Mercury's surface. While absorption bands are created in the near-IR spectral region from electronic transitions in the molecules bonded in the lattices of silicates, in this mid-IR region, the absorption bands are caused by the vibration, bending, and flexing modes of the crystalline lattices. In the mid-IR spectral region it is possible to measure the thermal emission from the regolith. When rocks and soils are warm they emit light that has a spectral signature of their composition. This is also true of molecules in the atmosphere. By measuring the spectral features from the thermal emission from a planet's surface and atmosphere we can learn what materials are there. The location on the planet, for all data, is biased toward the hottest regions in the footprint of the *spectrograph aperture*.

Methods used to interpret mid-IR spectra include the following:

(1) Identification of key spectral features diagnostic of composition. This has been achieved in the laboratory by measuring spectra from terrestrial and lunar surface materials like rocks, minerals, powders of rocks and minerals, and glassy volcanic material as well as glasses formed from rapid cooling of silicate melts.
(2) Comparison of laboratory and telescopic mid-IR spectra of lunar soils from similar locations on the Moon.
(3) Use of the same spectrograph to obtain spectra of rocks, minerals, and powders to compare to spectra of the Moon's and Mercury's surface and thus calibrate spectrograph performance and resulting spectral character.
(4) Comparison of spectra, obtained from spacecraft above Earth's atmosphere, of objects in the Solar System (Jupiter, Saturn, asteroids) with those obtained from mid-IR instruments from ground-based observatories.

8.3.6 Volume scattering region

From 2.5 to 7 μm, *volume scattering* of light from regoliths becomes important. There are absorption (and emittance) features associated with photons scattering in individual grains. Many silicates, sulfates, and carbonates have diagnostic features in this spectral region. Most of this spectral region is available to spectrographs at high-altitude ground-based observatories and from stratospheric observatories like the Kuiper Airborne Observatory (KAO – retired) and the Stratospheric Observatory for Infrared Astronomy (SOFIA – to be commissioned in 2005).

Spectra from olivine and pyroxene, two minerals common in igneous rocks, exhibit emission peaks in the volume scattering region. Specifically, pairs of peaks near 5.7 and 6.0 μm are diagnostic of olivine while a broad emission maximum near 5 μm is diagnostic of pyroxene.

A strong 5 μm emission feature in a Mercury spectrum from 45–85° longitude closely resembles that of laboratory clino-pyroxene powders. The best fit is to diopside ($CaMgSi_2O_6$), and the low-FeO abundance indicated by near-IR reflectance spectroscopy supports a low-iron-bearing pyroxene. Spectroscopic observations of Mercury have been made in this spectral region from the KAO, and between 3 and 7 μm from 13,786 ft altitude at Mauna Kea (the NASA Infrared Telescope Facility [IRTF]). Another emission feature, the emission maximum (EM), is also present in the spectrum. The wavelength of the EM is an important diagnostic of the weight percent of SiO_2 in the silicate. This will be discussed in Section 8.3.7. Figure 8.4

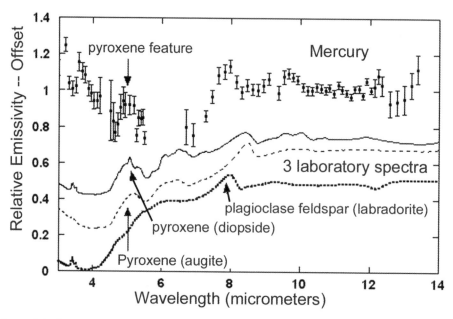

Figure 8.4. A spectral feature at 5 μm in Mercury's spectrum from 45–85° longitude, resembles that exhibited in laboratory spectra from two samples of pyroxene. Also exhibited is an emissivity maximum (EM) at 7.9 μm that is indicative of intermediate SiO_2 content. Plagioclase feldspar of ~Ab_{50} also has an EM at this wavelength. Spectra such as these indicate that Mercury's surface, at these locations, has spectral characteristics similar to those of low-iron basalt, or anorthosite with plagioclase more Na-rich than lunar anorthosite (laboratory spectra from Salisbury *et al.*, 1986, 1991; Mercury spectrum from Sprague *et al.*, 2002).

shows the Mercury spectrum from 45–85° longitude along with some laboratory spectra of pyroxene and plagioclase feldspar for comparison. Unfortunately the two pieces of information (pyroxene and weight % SiO_2) are not enough to uniquely identify the rocks that formed the regolith at this location. Powders of either low-iron basalt or anorthosite with about 90% plagioclase and 10% low-iron pyroxene could give the same features.

The recent reanalysis of *Mariner 10* imaging has also revealed new information about this region of Mercury's surface, especially around the region of Rudaki plains (3°S , 56°W) and Tolstoj (16°S, 164°W). Here smooth plains have *embayed boundaries* indicative of lava flow infills. Also scattering properties are similar to those of *pyroclastics* and glasses on the Moon. On the Moon, lava flows are basalt. On Mercury there is no evidence for the iron-bearing basalts so common on the Moon. However there is ample evidence for rocks formed from fluid lava flows. Such fluid lava flows could have the composition of low-iron basalts or other types of low viscosity lava (Figure 8.5).

8.3.7   Reststrahlen bands and emissivity maxima (EM)

Major rock-forming minerals have their fundamental molecular vibration bands in the region from 7.5 to 11 μm. These features are called the Reststrahlen bands. The transparency feature between 11 and 13 μm is associated with the change from surface to volume scattering. There are also major features in the region from 13 to 40 μm, associated with the bending, twisting modes of silicates, and other solar system materials.

Emissivity maxima (EM) are associated with the *principal Christiansen frequency*, a silicate spectrum which usually occurs between 7 and 9 μm. As mentioned above, EM are good diagnostics of bulk regolith or rock type (weight percentage SiO_2), in powdery mixtures of rocks, minerals, and glasses common in regoliths. The EM is also a spectral diagnostic of specific mineral identity.

On Mercury, EM at or close to 7.9–8.0 μm occur in spectra from 12–32°, 22–44°, 40–45°, 45–85°, 10–75°, and 110–120° longitude (Figure 8.6 and 8.7), indicative of intermediate silica content (~50–57% SiO_2). These locations are in the intercrater plains east of the crater Homer.

Based on the spectrum shown in Figure 8.7 and from other spectra of Mercury's surface, it may be that Mercury's regolith has a high concentration of plagioclase feldspar, perhaps labradorite $NaAlSi_3O_8$ $CaAl_2Si_2O_8$. An alternative explanation for the apparent match to plagioclase feldspar is that the spectrum comes from the glassy soil on Mercury's surface that is very mature after long periods of meteoritic bombardment. Scientists have shown that if lunar soils are very mature, much of the FeO is removed from the glasses and they appear much more feldspathic (like feldspar) in laboratory spectral measurements.

Many smaller dips in the Mercury spectrum are not present in the spectrum from the laboratory sample. These features may be in the Mercury spectrum because Mercury's surface is much hotter than the environment of the chamber holding the laboratory sample. Alternatively, the features may be from other minerals that

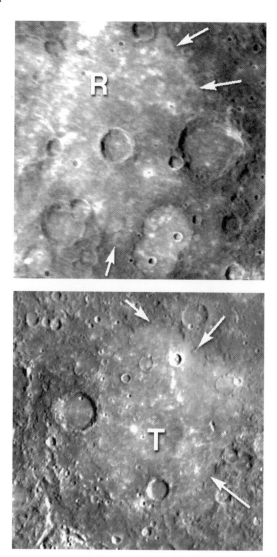

Figure 8.5. Spectra from equatorial regions at 45–85° longitude shown in Figure 8.4. come from the region on Mercury's surface shown in the top image. Rudaki (R) is a region where multi-spectral image analysis has revealed evidence for volcanic flow fronts and pyroclastics. In the lower image of Tolstoj (T) regions of embayments from apparent lava flows are also indicated with arrows. (adapted from Robinson and Taylor, 2001).

were not in the model. It is also possible that some of the smaller features are noise in the Mercury spectrum and do not represent spectral features from materials on Mercury's surface. Spectra from 68–108° and 100–160° longitude have multiple EM indicating a more complicated bulk composition and/or mixed mineralogy of lower SiO_2 content (more (mafic) basic composition, 45–49% SiO_2).

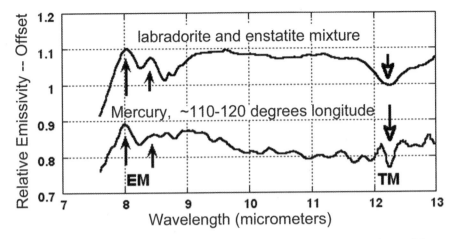

Figure 8.6. Shows an example of one of these Mercury spectra compared to a laboratory spectrum from a mixture of plagioclase feldspar and the Mg-rich ortho-pyroxene called enstatite. The EM and the transparency minima (TM) are marked with arrows. Spectra obtained in the laboratory from powdered samples of silicates and alumino silicates are often found to be similar to spectra from Mercury. The wavelength and the shape of the EM at 8.0 µm is the same for the Mercury spectrum and the model spectrum. The wavelength of the TM for Mercury the model also matches. Differences between the two spectra in the Reststrahlen bands can also be seen but are not yet explained. Adapted from Sprague and Roush (1998).

8.3.8   Transparency minima (TM)

The TM between 11 and 13 µm is associated with the change from surface to volume scattering. Generally the wavelength where the minimum occurs is a good indicator of the weight percent SiO_2 present in a rock powder. Because of historical geologic circumstance, the divisions of SiO_2 are designated by the terms acidic, intermediate, basic, and ultrabasic. The range of wavelengths for transmission minima and the weight percent SiO_2 content for some common rock powders are shown in Table 8.3.

The classification of silicates according to weight % SiO_2 has been replaced in some cases by classification according to mineralogy. For the discussion in this chapter we prefer the older nomenclature but include the newer paranthetically in Table 8.3 and point out that the SiO_2 values are appropriate for typical samples.

The Mercury spectrum from 110–120° longitude in Figure 8.6. has a clear and strong transparency minimum at 12.3 µm that is at the same location as the transparency minimum in a laboratory spectrum of labradorite powders or some low-iron basalt. This is consistent with the location of the EM in the same spectrum as described above. Spectra from longitudes centered on 80°, 256°, and 266° have probable transparency minima at 12 µm. The bulk composition associated with a transparency feature at this wavelength is intermediate to basic (45–57 weight % SiO_2). Spectra from a region centered on 15° has a minimum at 12.5 µm indicative of about 44 weight % SiO_2 or an ultra-basic composition. A spectrum from a region

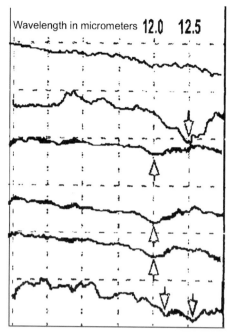

Figure 8.7. Spectra from Mercury's surface at 6 different locations, show TM from top down at longitudes centered on: 10°, 15°, 256°, 266°, 80°, and 229° respectively. Wavelengths are between 12 and 12.7 µm. Such spectral features are indicative of the weight % SiO_2 in the soil at the location measured. Generally, the shorter the wavelength of the TM, the more SiO_2 in the silicate (see text for more details) (adapted from Cooper *et al.*, 2002).

Table 8.3. Wavelength ranges for TM and weight percentage SiO_2 for common rock powders.

Designation	Weight % SiO_2	Rock powder (intrusive)	Rock powder (extrusive)	Wavelength (µm) of TM
Acidic (felsic)	62–75	Granite	Rhyolite	11.5–11.7
Intermediate	52–61	Diorite	Andesite	11.8–12.1
Basic (mafic)	46–51	Gabbro	Basalt	12.1–12.4
Ultrabasic (ultramafic)	39–45	Peridotite	Picrite	12.4–12.7

centered on 229° longitude has a doublet transparency minimum. Figure 8.7 illustrates these spectral features with 6 spectra from Mercury's surface obtained with the McMath Pierce Solar Telescope near Tucson, Arizona.

About 40% of Mercury's surface has been measured spectroscopically. Roughly speaking the coverage is of the equatorial and low latitude regions at most, but not all, longitudes. From these observations we know that Mercury's surface composition is heterogeneous. Regions near Homer and the Murasaki crater complex appear feldspathic, trending toward Ab_{70}–Ab_{40}, more Na-rich than the lunar anorthosites.

Bulk compositions are of intermediate silica content. Some mixed compositions of lower silica content are present in the regions from 68–160° longitude, but not at all locations measured. One candidate for the rocks at these locations is low-iron basalt – mixtures of feldspar, pyroxene, and minor olivine of intermediate (52–57 weight percent) or basic (46–51 weight percent) SiO_2.

Regions west of Caloris appear to have mixed mineralogy and a more complex bulk composition with some basic and ultrabasic regolith types. According to transparency minima, an ultrabasic composition falls east of the Homer crater near 15°. Two measurements indicate that ultrabasic regolith types are located far west of Caloris from about 205° longitude. Feldspathoids, alkali-rich alumino silicates with low (39–45 weight percent) SiO_2, such as tephrite or basanite are possible extrusive rocks that could be in this region. Such rocks are formed from plagioclase bearing lavas in which feldspathoids are present in greater abundance than 10 weight percent. Tephrites have little olivine while basanites may have considerable olivine. There is evidence for ultrabasic-like soils at 205–240° longitude. Peridotite and dunite are olivine-rich low SiO_2 rocks. It would not be surprising to have low-iron olivine in these regions but no signature identified as olivine has yet been found.

This interpretation is very generalized because the areal extent of the spatial footprint is no smaller than 200 km by 200 km for the very best spatially resolved observations, and as much as 1000 km by 1000 km for the least spatially resolved region. On the surface of Mercury viewed by *Mariner 10* these large areas mostly encompass a variety of morphologically different terrains that could have different compositions.

8.3.9   Comparison to the Moon

Hundreds of kg of lunar samples were brought back by *Apollo* astronauts. Some of those samples have been measured in the laboratory with spectrometers to examine and compare their spectral signatures from location to location.

A spectrum from equatorial regions near 20–25° longitude on Mercury's surface is shown and compared to laboratory spectra from a *particulate breccia* sample brought back from the *Apollo 16* landing site on the Moon (Figure 8.8). The lunar sample is ~90% anorthite (Ca-feldspar) and ~10% pyroxene. The EM for both spectra is centered close to 8 μm and marked with an arrow. The wavelength of the EM indicates a *feldspathic* rock type. Other features in the Mercury spectrum are also similar to the lunar spectrum and indicate the presence of pyroxene on Mercury's surface.

Based on these observations and others in the near-IR region of the spectrum, some scientists who specialize in the petrology of the lunar surface think that lunar anorthosites provide the best analogue for the composition of Mercury. Also there are examples of meteorites with little or no oxidized iron such as the *enstatite chondrites* and *achondrites*. These meteorites formed in a *reducing* environment and have sulfides present. It has been suggested that Mercury may have formed in a similar environment. One argument against this theory is that mid-infrared spectra

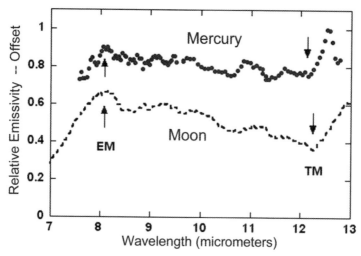

Figure 8.8. A spectrum from Mercury's equatorial regions at about 20–25° longitude is shown along with a laboratory spectrum of a sample from the Moon. The lunar sample is a particulate breccia #67031 (composed of feldspar and pyroxene) from the *Apollo 16* landing site (mercury spectrum from Sprague *et al.*, 1996; lunar spectrum from Nash and Salisbury, 1991).

from Mercury's surface do not resemble laboratory mid-infrared spectra from such meteorites. Of course as more data become available, this position may change.

8.4 WHERE IS THE IRON AT MERCURY?

Mercury has a large, high density core which is probably composed of mostly iron with the addition of some lighter elements such as sulfur. Issues regarding Mercury's core were partially discussed in Chapter 5 and will be discussed in more detail in Chapter 12. As discussed above, there is little evidence for FeO on Mercury's surface in any data set. This is certainly one of the most interesting puzzles about Mercury. Has all of Mercury's iron gone into the core and left none behind in the regolith, crustal rocks, or mantle?

8.4.1 Other terrestrial planets and the asteroid belt abound with oxidized iron

Venus, Earth, the Moon, Mars, and many asteroids have several percent oxidized iron in their surface rocks and minerals. The abundance of oxidized iron (FeO or Fe_2O_3), for example, has been measured or inferred in a variety of ways – no one way is suitable for every rocky body. For Venus, the *Venera 13* and *14* landers made *in situ* fluorescent X-ray emission measurements of the soil and cores from a drilling experiment. At the *Venera 13* site a high potassium basalt was discovered that is similar to mid-oceanic ridge basalts. At the *Venera 14* site the rock composition is

Table 8.4. Abundance of FeO in surface rocks and regolith inferred from measurements of terrestrial planets, the Moon, and some asteroids.

Terrestrial Body	Typical weight % FeO
Mercury	1–3
Venus	8–10
Earth	5–28
Moon	4–27
Mars	10–40
Asteroids	4–25

similar to plateau basalts on Earth. These basalts have an FeO content of from 8–10 weight percent. The Earth's surface basalts have an FeO content ranging from <2 to 28 weight percent. For the Moon, remote sensing from Earth-based telescopes and returned lunar samples from the *Apollo* and *Luna* missions show 4–27 weight percent FeO. Meteorites from Mars, and remote sensing from Earth-based telescopes and orbiters indicate a FeO content of 10–40 weight percent. For asteroids, near-IR spectroscopy has revealed a wide variability in iron content ranging from 4 to 25 weight percent. But almost all asteroids in the asteroid belt show some hint of the famous reflectance absorption features that indicate oxidized iron in pyroxene or olivine, or both. A comparison of the abundance of FeO in surface rocks and regolith inferred from many measurements of all the terrestrial planets, the Moon, and some asteroids is given in Table 8.4. Iron content in meteorites ranges from almost all to practically none.

8.4.2   For Mercury, low oxidized iron

All the telescopic spectral evidence (near-IR, mid-IR, and radio wavelengths) for Mercury suggests that its surface is low in oxidized iron. This suggests that Mercury is a chemically unique terrestrial planet. This also places severe constrains on Mercury's origin as discussed in Chapter 12. These measurements indicate that the FeO content is from 1–3 % at most. As mentioned above, it could be that space weathering has converted most of the FeO to iron metal but this explanation fails to meet the constraints of the radio centimeter wavelength observations that show Mercury's regolith to be more transparent (lower in iron and iron metal and titanium) than the Moon.

Recalibrated *Mariner 10* images taken in the UV (375 nm) and orange (575 nm) wavelengths indicate compositional variations in the Mercurian surface consistent with those deduced by earth-based spectroscopic observations. These newly calibrated color images have been interpreted according to the view that ferrous iron lowers the albedo and reddens a lunar or Mercurian soil. These data suggest that the average hemispheric FeO content is less than 3% by weight, which is consistent with Earth-based measurements. Furthermore, the color images show color boundaries between smooth plains and the surrounding terrain indicating compositional differences. In at least two cases the smooth plains overlie material that is bluer (higher UV/orange ratio) and is enriched in opaque minerals relative to the hemispheric average. Since the smooth plains are probably old lava flows (see Chapter 10), and the FeO solid/liquid *distribution coefficient* is about 1% during partial melting, it is estimated that the Mercury's mantle has a FeO abundance similar to the lava flows (less than 3%). But this is very sensitive to the degree of partial melting.

8.5   SUMMARY

Both Earth-based spectroscopic observations and calibrated *Mariner 10* images indicate that the surface of Mercury is heterogeneous in composition with a wide range of SiO_2 content. The FeO content appears to be between 1 and 3% which is abnormally low compared to other terrestrial planets and the Moon. Evidence for pyroxene appears to be of the Mg-rich or Ca-rich type. The spectroscopic data are consistent with compositions ranging from low-iron basalts and anorthosites. There are also spectra that exhibit similarities to laboratory spectra of *syenite*. However, to have a rock so highly evolved petrologically requires multiple episodes of partial melting which may be problematical for Mercury.

Photometry of Mercury's surface in the UV and visible indicates Mercury is fairly smooth, consistent with flooding by lavas. The morphology of the land forms, which will be discussed in detail later in Chapter 10, indicate fluid lava flows over much of the surface. *Mariner 10* imaging ratios indicate bright excavated regions. Ground-based spectroscopy indicates that these excavated regions may be anorthosites. This would be consistent with the appearance of enhanced regions of Na atmosphere that are associated with the fresh craters as discussed in Chapter 6. Continued ground-based observations and detailed measurements by *MESSENGER* should greatly expand our knowledge of the variety of compositions and their spatial distribution.

9

The impact cratering record

9.1 MERCURY'S MOST COMMON LANDFORM

Mercury is one of the most heavily cratered planets in the Solar System, and its cratering record provides important information on the cratering process and crater characteristics in that part of the Solar System. Because Mercury is the innermost planet, it provides important constraints on the origin of impacting objects in the terrestrial planet domain.

9.1.1 It all began with the Moon

In 1609 Galileo recognized and wrote about craters he viewed with the recently invented telescope. The craters he saw, were common on the Moon. In fact, the most common landforms in the Solar System are impact craters. They occur in greater or lesser abundance on almost all solid bodies explored to date.

9.1.2 Three basic crater characteristics

There are three basic characteristics common to all relatively fresh impact craters:

(1) a near-circular raised rim;
(2) a floor that is deeper than the crater surroundings; and
(3) a relatively rough *ejecta blanket* that surrounds the crater.

Small craters have bowl-shaped interiors and are called simple craters. Larger craters have terraced inner walls, a relatively flat floor, *central peaks* and are called complex craters. The rim structure consists of a flap of overturned material resulting in inverted *stratigraphy* (older on top and younger on the bottom). The crater is surrounded by an extensive ejecta deposit consisting of two parts: a relatively narrow inner zone of continuous hummocky ejecta; and an outer zone consisting of strings and clusters of secondary craters caused by the impact of discrete masses of ejecta. Fresh craters have ray systems consisting of newly excavated material associated with secondary craters.

On Mercury, impact craters are found on all types of terrain and in various states of preservation. They are the dominant landform on the planet. The largest relatively well preserved impact feature seen by *Mariner 10* is the 1300 km diameter Caloris basin. Probably craters less than a millimeter in diameter have formed from dust-sized micrometeorites, based on *Apollo* lunar returned samples. Some craters are fresh with extensive *ray systems* while others are so degraded that only discontinuous remnants of their rims remain. It is believed that the two large radar anomalies (A and B) on Mercury's unseen side are relatively recent impact craters (see Chapter 7).

9.2   CRATER FORMATION

9.2.1   Energy of impact

When high velocity objects strike planetary surfaces they produce enormous amounts of kinetic energy. The amount of energy produced is $\frac{1}{2}mv^2$ where m is the mass of the object and v is its impact velocity. For example, if a 1-km diameter iron meteorite hit the Earth at 15 km/s it would release an amount of energy equivalent to about 100,000 megatons of TNT (1 megaton is 1 million tons). That amount of energy would produce a crater about 12 km in diameter.

9.2.2   Crater diameter and depth

The diameter of impact craters depends on a number of parameters besides velocity and mass. Among these are the size of the projectile, the ratio of projectile to surface density, surface gravity, impact angle, and for larger *complex craters*, the transition diameter from simple to complex craters. Because Mercury is so close to the Sun, the large gravitational pull of the Sun causes objects to impact Mercury at velocities greater than all other planets for given projectile orbital characteristics. For instance, on average, asteroids will impact Mercury at a velocity of about 34 km/s, compared to 22 km/s on the Moon and 19 km/s on Mars. Parabolic comet impacts (comets from the outer fringe of the Solar System) should be much more frequent on Mercury than other bodies (about 41% of the craters on Mercury, about 10% on the Moon and Earth, and less than 3% on Mars). On Mercury, comet impacts will have an average velocity of about 87 km/s compared to 52 km/s on the Moon and 42 km/s on Mars. Therefore, craters will generally be larger and produce more melt and ejecta on Mercury than on other planets and satellites for similar sized objects with similar physical characteristics.

In an impact event, kinetic energy is rapidly transferred to the planetary crust. Most of the energy takes the from of shock waves that travel at supersonic speeds through both the crust and the impacting object. They spread out in a hemispherical shell from the point of impact. As the shock waves pass through the rocks they are subjected to very high pressures that can rise to hundreds of kilobars (kbar). Granite is crushed at 250 kbar, melted at about 450 kbar and vaporized at 600 kbar. It is the interaction of the shock waves with the unconfined surface that excavates the crater. As the shock wave passes through the compressed rocks they snap back along the unconfined surface. This produces what is called a tensional *rarefaction wave* that decompresses and fractures the rock. The net effect is to momentarily convert the rock to a fluid-like material that moves laterally upward and out of a steadily growing excavation cavity. The cone of rapidly moving ejecta is mainly deposited beyond the crater's final rim. The crater stops growing when the strength of the target material exceeds the decaying strength of the shock wave. This initial crater is called the excavation crater, but may be enlarged by slumping of the rim into the crater if the crater is large enough. The rock layers at the edge of the crater are pushed upward and overturned by the passage of the shock wave to produce the characteristic raised rim of impact craters. Of course, well before this time the projectile has been completely destroyed as the shock waves generated in the pro-jectile interact with the unconfined surface. It essentially explodes, some of it vaporized, some melted, and the rest shattered into small pieces. It is possible to see many of these features of impact craters on Earth, for example, at Meteor crater near Winslow, Arizona. It was formed when an iron meteorite estimated to have been about 30 m in diameter and with a mass of about 100,000 tons excavated a crater about 1.2 km in diameter. The eroded rim rises about 47 m above the sur-rounding topography, and 174 m above the crater floor.

9.2.3   Volatilization and melting of surface and impactor

Not all the kinetic energy of the impactor is used to excavate the crater. Some is partitioned into heat. The heat can be so great that a large volume of the target material is melted and volatilized. In large craters impact melt is found as a sheet overlying fragmented floor material, as ponds and flows on the crater rim, and as part of the continuous ejecta blanket. Great plumes of atoms and molecules may be sent far above the surface and contribute to a temporary atmosphere.

As discussed in Chapter 6, much of the Na, K, and Ca atmosphere observed from ground-based spectroscopy is created in the volatilization following impact of interplanetary dust particles on Mercury's surface.

9.3   CRATER MORPHOLOGY

9.3.1   Three general crater morphologies

The morphology of Mercurian craters is similar to that of lunar craters in most respects. Like the Moon, the general structure of Mercurian impact craters can be divided into three types (Figure 9.1). At diameters less than 10 km they have bowl-shaped profiles with raised rims. At diameters greater than 10 km they have central peaks, flat floors and terraced inner walls. Therefore, on Mercury the transition

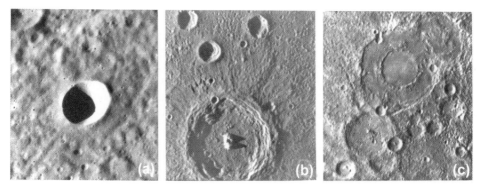

Figure 9.1. Three Mercurian craters show the change in morphology with increasing size. In (a) the crater is 8 km diameter and has a bowl-shaped depression. At diameters between 15 and 100 km the craters have central peaks and terraces on the interior rims. The crater Brahms in the center (b) is a complex crater 75 km diameter. At diameters greater than 100 km craters develop central rings as shown by the 225 km Bach basin (c).

diameter from simple to complex craters occurs at 10 km. The transition diameter is not the same on all bodies. On the Moon it occurs at 19 km and on Earth it is about 3 km. The transition diameter depends primarily on the surface gravity; the stronger the surface gravity the smaller the transition diameter. Although gravity seems to be the most important factor controlling the transition diameter, the physical character-istics of the target material are also important. For example, on Earth the transition diameter is smaller in weaker sedimentary rocks than in stronger crystalline rocks. On Mars the transition diameter is smaller than on Mercury (5 compared to 10 km) although their surface gravities are the same. This has been used as evidence for a weaker ice/water-rich layer on Mars.

In small *simple craters* there is little or no inward slumping of the rim, and the final crater is essentially the excavation crater. In complex craters the excavation crater is enlarged by inward slumping of the rim. With large craters between 15 to 200 km diameter rim slumping can enlarge the crater considerably. At very large diameters associated with impact basins, whole sections of the crust collapse into the excavation cavity to enlarge the diameter by tens to hundreds of kms.

At the largest diameters the craters show double or multiple rings (Figure 9.1(c)). At these sizes they are usually referred to as impact basins. On Mercury double ring basins begin to form at about 200 km diameter and multiple rings at about 750 km diameter. At the lower diameters of double ring basins, central peaks are usually present. The morphology of these impact basins will be discussed in more detail in Section 9.5.

9.3.2   Difference in Physical Properties of Lunar and Mercurian Highlands

There are significant differences in the abundances of central peaks, terraces, and scalloped crater rims between fresh craters in the lunar maria and highlands

Colour plates

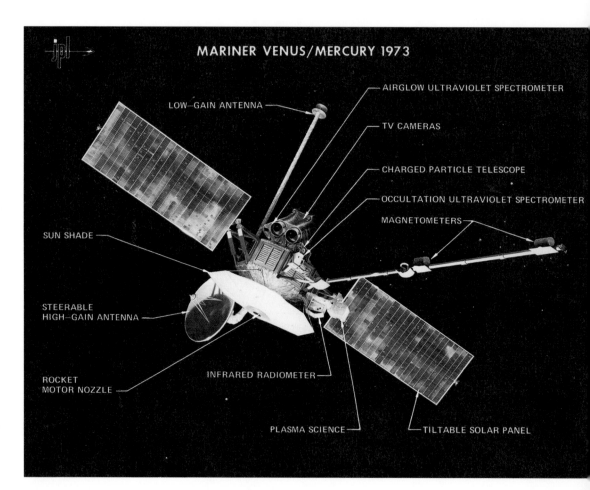

Figure 2.2. This illustration of *Mariner 10* shows the solar panels, long magnetometer boom, the sun shield, the high-gain antenna, and the science instruments. The pole-like object at the top is the omnidirectional, low-gain antenna (courtesy of Jet Propulsion Lab., Pasadena, CA).

Figure 2.3. At the moment the launch window opened, *Mariner 10* was sent on its way to Mercury on 3 November, 1973 at 12:45 a.m. Eastern Standard Time.

Figure 2.7. The first flyby of Mercury took place on the planet's night side. This illustration shows *Mariner 10*'s flightpath by Mercury and some of the image footprints and infrared traces taken on the planet's surface (courtesy of Jet Propulsion Laboratory, Pasadena, CA).

Figure 2.10. The second flyby on Mercury's day side permitted the acquisition of images of the southern hemisphere which joined the two sides seen on the first flyby (see Figure 2.12) (courtesy of Jet Propulsion Laboratory, Pasadena, CA).

Figure 4.8. *Mariner 10* photomosaic of the incoming side together with an accurate artist's rendition of the size of Mercury's core compared to the silicate portion. The outer part of the core may still be in a liquid state (from Strom, 1987).

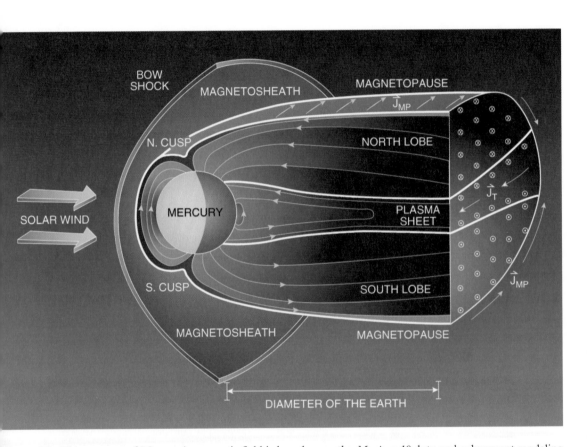

Figure 5.2. This concept of Mercury's magnetic field is based upon the *Mariner 10* data and subsequent modeling and analyses of those data. It is possible that this view will change with new observations made by particles and fields instrumentation on future missions to Mercury (courtesy of James Slavin, NASA Goddard Space Flight Center, Laboratory for Extraterrestrial Physics, Greenbelt, Maryland, USA).

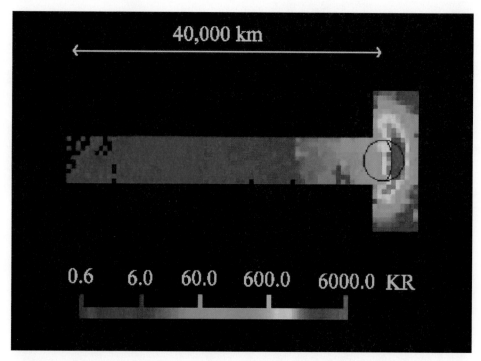

Figure 6.2. Images of the sodium emission intensity in kilo-Rayleighs (KR) measured anti-sunward of Mercury at 0300 UT 26 May, 2001. Each square represents an observation with a 10×10 arcsec image slicer, and represents an area 5100 km. The emission intensity varies over four orders of magnitude, decreasing from about 6000 KR at the planet to about 0.6 KR at the extreme end of the tail. The T shape is not part of the shape of the Na cloud (courtesy of Andrew Potter).

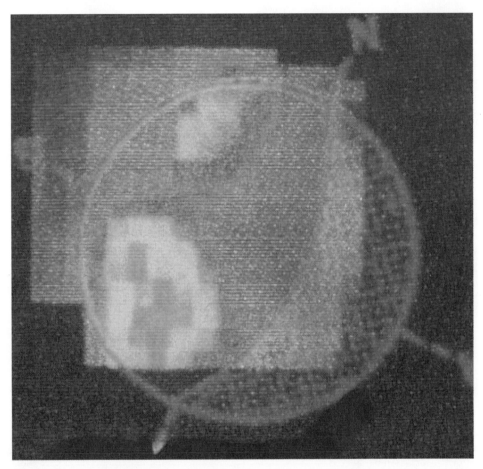

Figure 6.3. Images of Mercury's atmospheric Na emission show bright spots. Some scientists think these are spots where ions directed toward the surface by electric fields near Mercury have sputtered neutral Na off the surface. Other scientists believe the Na is bright at these locations because there are freshly exposed Na rich rocks and soils at those regions and that the Na is released when the regolith is heated (adapted from Potter and Morgan, 1990).

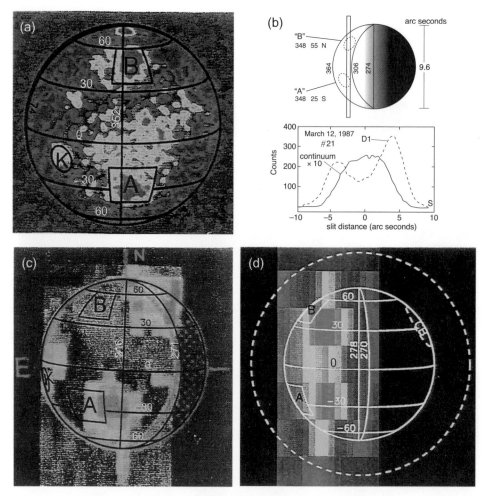

A + B = Na spots and radar bright regions
K = Kuiper Muraski crater
CB = Caloris basin

Figure 6.4. (a) Radar image of spots A and B (from Slade *et al.*, 1992). (b) Diagram of geometry for slit spectroscopy measurements of Na enhancements over spots A and B (from Sprague *et al.*, 1997). (c) Na enhancements over spot K, A, and an unidentified region (adapted from Potter and Morgan 1990). (d) Na enhancements from spectroscopic measurements interpolated to make an "image" (from Sprague *et al.*, 1998). (e) Another case of Na enhancements over K, combined with B and A (adapted from Potter and Morgan, 1990). (f) Spectroscopic measurements when the slit was placed over Caloris basin and an unidentified source (after Sprague *et al.*, 1997). (g) Na enhancement over Caloris basin (from Potter and Morgan, 1990). (h) Na enhancements over the regions of bright albedo features at: 155°W, 65°N; 125°W, 0; and 105°W, 9°S longitude and latitude respectively (from Potter and Morgan, 1990). (i) Enhancement of Na spreading from ∼43–73°W longitude and ∼10–60°S latitude (from Potter and Morgan, 1997).

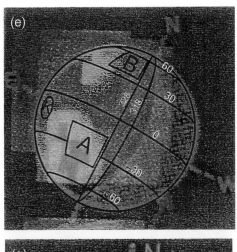

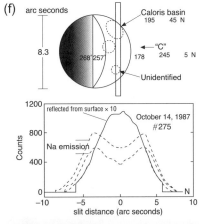

(f) arc seconds

Caloris basin
195   45 N

8.3

268" 257   178   "C"   245   5 N

Unidentified

October 14, 1987
#275

reflected from surface × 10

Na emission

Counts

1200

800

400

0

N

−10   −5   0   5   10
slit distance (arc seconds)

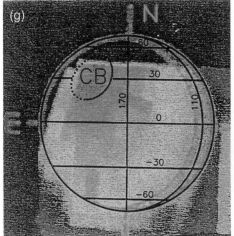

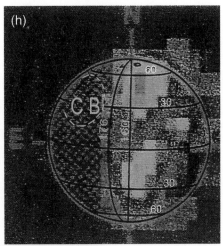

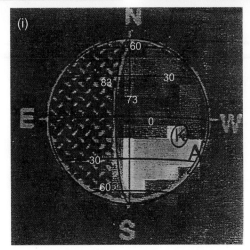

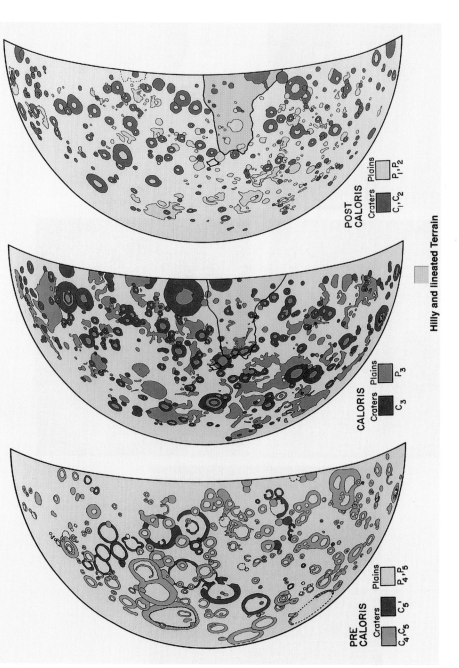

Figure 10.4. Paleogeologic maps of Mercury's incoming side seen by *Mariner 10* showing the distribution of craters and plains of various relative ages. The oldest craters (C_4–C_5) and plains (P_4–P_5) pre-date the hilly and lineated terrain shown in green and are probably pre-Caloris impact age. The C_5 craters predate the P_4–P_5 plains. The youngest craters (C_1–C_2) and plains (P_1–P_2) post-date the hilly and lineated terrain and are probably post-Caloris impact age. The P_5–P_3 plains are equivalent to intercrater plains and the P_2–P_1 plains are smooth plains (from Leake, 1981).

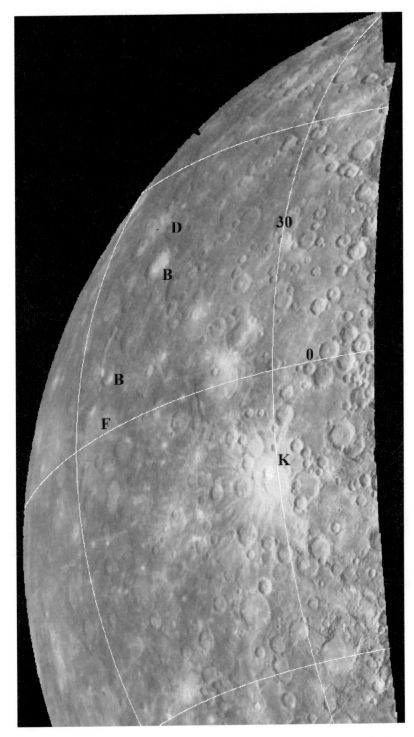

Figure 10.10. This false color photomosaic by Mark Robinson shows sharp color boundaries that coincide with geologic boundaries. They probably represent differences in composition. The Kuiper/Muraski crater complex is indicated by "K". The "F" and "B" indicates plains units with sharp color boundaries. The "B"'s are plains flooded craters. The largest is Lermontov (160 km diameter). The "blue" area at "D" may be material ejected from the subsurface by a fresh impact crater. See text for explanation (courtesy of Mark Robinson, Northwestern University).

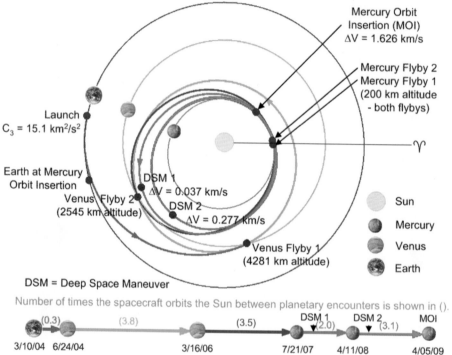

Figure 13.1. This diagram shows the trajectory of the *MESSENGER* spacecraft as it flies by Venus and Mercury before Mercury orbit insertion on 5 April, 2009. The Deep Space Maneuvers (DSM) are engine burns to correct the trajectory of the spacecraft. The ΔV is the change in velocity required by that maneuver. At the bottom of the diagram is a schematic timeline of the mission up to Mercury orbit insertion with the number of spacecraft solar orbits in parentheses (courtesy of James McAdams, Applied Physics Laboratory, Laurel, MD).

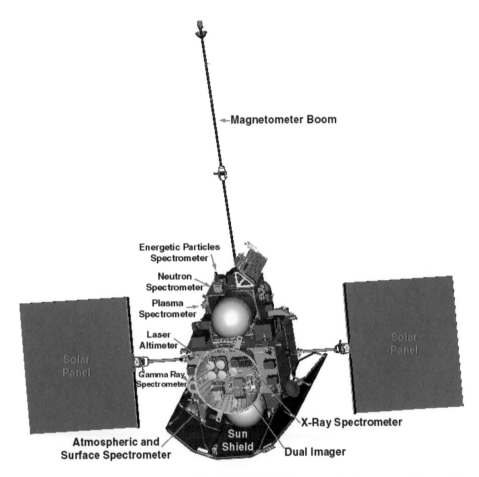

Figure 13.3. Schematic illustration of the *MESSENGER* spacecraft showing positions of the science instruments and other features (courtesy of Ralph McNutt, Applied Physics Laboratory, Laurel, MD).

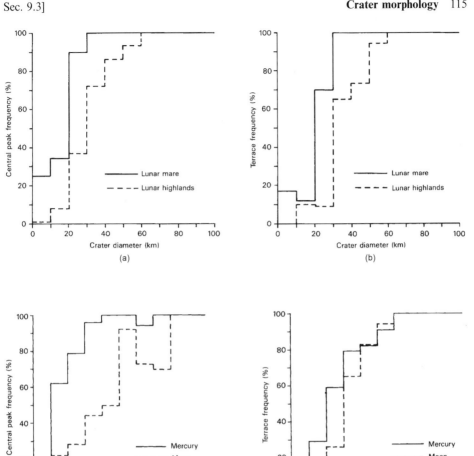

Figure 9.2. Histogram of the central peak (a) and terrace (b) frequency versus crater diameter for the lunar highlands and maria (top). Histogram of the central peak (a) and terrace (b) frequency versus crater diameter for the Moon and Mercury (bottom) (from Smith and Hartnell, 1979).

(Figures 9.2 and 9.3). This has been attributed to differences in the physical properties of the lunar highlands and maria. The highlands is composed of a thick regolith and breccia (the *megaregolith*), and the maria consist of a thin regolith underlain by relatively unbrecciated volcanic lava flows. Furthermore, the morphologies of craters formed in the lunar maria, the Mercurian smooth plains, and the Mercurian highland cratered terrain are similar. However, there are large differences between the crater morphologies in the lunar highlands and the analogous Mercurian

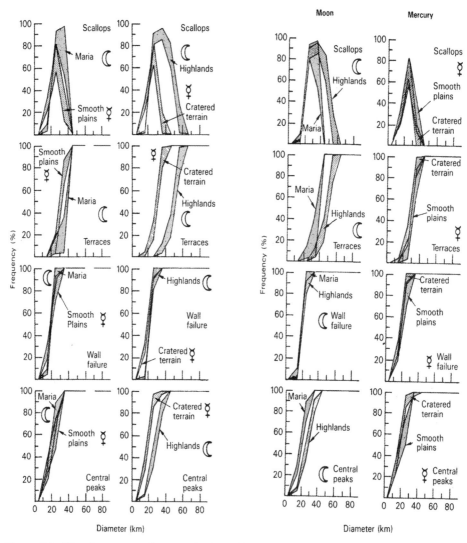

Figure 9.3. The first two columns of plots show the morphology/frequency distribution comparing craters on the lunar Maria to those in the Mercurian smooth plains (first column on left), and craters in the lunar highlands to those in the Mercurian cratered terrain (second column on left). The third and fourth columns are plots showing the morphology/frequency distribution, illustrated as ±1 sigma envelopes around the mean values, for craters on the Moon (third column) and Mercury (fourth column) (from Cintala *et al.*, 1977).

cratered highlands (Figure 9.3). This suggests that a difference in the physical properties of the target material, rather than surface gravity, is the major factor affecting interior crater morphology. The main difference between the lunar and Mercurian highlands is the great abundance of intercrater plains in the Mercurian highlands (see Chapter 10). The lunar highlands have only small patches of intercrater plains that can be identified as ejecta deposits from certain basins (see Figure 9.21). These differences suggest that the Mercurian intercrater plains consist of a more coherent, stronger material akin to solid rock, rather than the less coherent, weaker mega-breccia of the lunar highlands.

9.4 EJECTA DEPOSITS

9.4.1 Two distinct regions of crater ejecta

Impact crater ejecta consists of two parts:

(1) a continuous ejecta blanket; and
(2) discontinuous ejecta beyond the continuous ejecta.

The continuous ejecta consists of a blanket of hummocky ejecta extending about 0.5 to 1 crater diameter from the crater rim. The area covered by the blanket can be four to nine times the area of the crater. Discontinuous ejecta consists of swarms of secondary impact craters formed by clots, strings, or individual fragments thrown beyond the continuous ejecta blanket. Individual fragments can be thrown for hundreds or thousands of kilometers. Very fresh craters have bright ray systems that consist of secondary craters having their own ejecta deposits consisting of fresh, bright material. Powdery material created by the impact also contributes to the rays. Ejecta can have far-reaching effects on planetary and satellite surfaces. If fragments are ejected at velocities exceeding the escape velocity of the planet or satellite, they will not return to the surface. In fact, we have samples of Mars and the Moon here on Earth that were ejected at velocities greater than the escape velocities of the parent bodies.

Ejected particles travel on looping paths called ballistic trajectories. For airless bodies like the Moon and Mercury, the distance a fragment will travel depends on the velocity and angle at which it is ejected and the gravity field of the planet. Most ejecta material is ejected at angles between 30° and 50° from the horizontal. For any given ejection velocity, a fragment will travel farther when ejected at an angle of 45° (Figure 9.4). Another parameter that effects the distance traveled is the radius of curvature of the planet or satellite: the smaller the radius of curvature the greater the distance traveled. On bodies with an atmosphere like the Earth and Venus, atmospheric drag will reduce the distance ejecta can travel.

9.4.2 Ejecta differences between Mercury and the Moon

The characteristics of Mercurian ejecta deposits are different from those on the Moon. On Mercury the continuous ejecta deposits and secondary impact craters are closer to the crater rim than similar sized craters on the Moon. This is the result of the greater surface gravity on Mercury ($370 \, \text{cm/sec}^2$) than on the Moon ($162 \, \text{cm/sec}^2$); for a given ejection velocity objects will travel about half the distance on Mercury. Thus the continuous ejecta blanket only extends outward to about

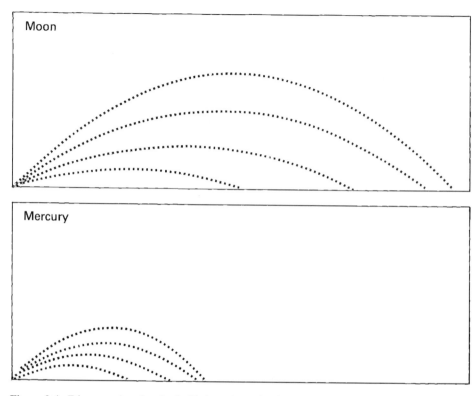

Figure 9.4. Diagram showing the ballistic trajectories for ejecta on the Moon and Mercury for the same impact conditions. Because of Mercury's much stronger gravity field, ejecta will travel more than twice as far on the Moon than on Mercury (from Strom, 1987).

0.5 crater radius. Also individual fragments travel shorter distances on Mercury than the Moon. On Mercury strings of secondaries often occur on the continuous ejecta blanket very close to the rim. This is rarely the case on the Moon. Thus, on Mercury both the continuous ejecta deposit and a greater abundance of secondary craters are concentrated nearer the crater rim (Figures 9.5 and 9.6). One apparent contradiction to this is the observation that some fresh craters have individual rays that extend enormous distances: much greater than fresh lunar craters of comparable size. Possibly these craters were formed by parabolic comets whose impact velocity at Mercury is exceedingly high (see section 9.2 on impact velocities on Mercury). In these cases possibly the ejection velocity of some swarms of fragments was extremely high and made up for the greater gravity field.

9.4.3   Crater degradation

The formation of impact craters and their ejecta deposits takes only seconds to minutes depending on the size of the impact. Over time, however, craters are

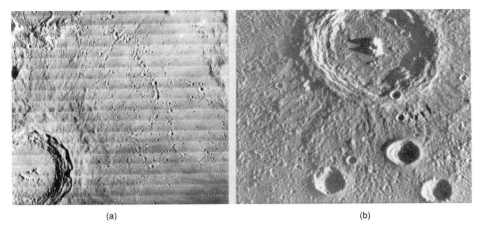

(a)   (b)

Figure 9.5. Comparison of the ejecta deposits for the lunar crater Copernicus (a) and a similar sized Mercurian crater (b). The ejecta deposit on the Mercurian crater is closer to the rim because of the higher gravity.

modified by a variety of processes. On Mercury, craters have been modified by three processes:

(1) subsequent impacts by both large and small objects including ejecta;
(2) volcanic deposition; and
(3) tectonic deformation.

Subsequent impacts have destroyed portions of pre-existing rims and ejecta blankets have obscured crater structures. Volcanic flooding has obliterated ejecta blankets or partially buried craters, and tectonic deformation has shifted the rims or floors and distorted the shape of craters. These processes have resulted in various degrees of crater degradation from slightly modified rims and ejecta deposits to barely discernable discontinuous rims (Figure 9.7).

9.5   THE CALORIS AND OTHER IMPACT BASINS

Very large impacts that form basins are devastating events for a planet or satellite. Their effects are so widespread that few areas of the planet are unaffected. Large impacts can trigger internal events that affect large areas of the planet. About 16 multiple ring basins larger than about 250 km diameter have been recognized on the 45 percent of Mercury observed by *Mariner 10*. About 15 impact basins have been located on the entire Moon. The largest basin so far seen on Mercury is the 1,560 km Borealis basin located near the north pole. This basin is very old and has been severely degraded. It is filled with smooth plains that embay and partially cover older craters. Much of the 45 percent of the planet was seen at high sun angles where it is difficult to discern structures, so it is possible that other basins occur in this 45 percent of the planet.

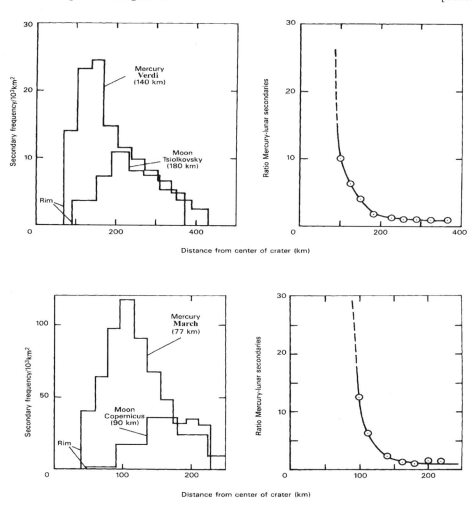

Figure 9.6. Plot of the radial variations in the areal density of secondary craters and the ratios of Mercurian to lunar secondary craters for the Mercurian craters Verdi and March and for the lunar craters Copernicus and Tsiolkovsky (from Gault *et al.*, 1975).

On Mercury the inner rings of the basin are often low, partial, or discontinuous, and, therefore, more inconspicuous than those on the Moon. Unlike the Moon, basins commonly have a partial, weak ring exterior to the main ring. The radial spacing of interior rings increases incrementally outward by about $\sqrt{2}$ of the diameter.

The basin topography and transient cavity size of Beethoven basin has been compared with lunar and Martian basins in conjunction with gravity data on those bodies, to infer the crustal viscosity at the time of impact. This comparison suggests that the crustal viscosity of Mercury during the period of heavy bombardment was relatively high, possibly due to an extremely dry crust.

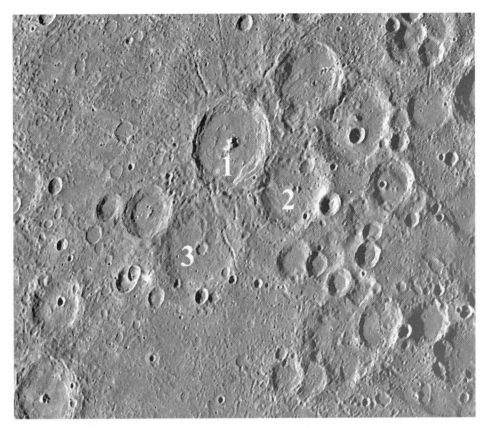

Figure 9.7. This image in the heavily cratered highlands of Mercury shows the various degradational stages of craters. Crater 1 is the freshest crater with a sharp rim and prominent ejecta deposits, while crater 2 has been degraded by subsequent cratering and the ejecta deposit of crater 1. Crater 3 is even more degraded by subsequent impacts, secondary cratering, and the flooding of its southern rim by intercrater plains material.

The largest, best preserved impact feature observed by *Mariner 10* is the Caloris basin (Figure 9.8). It is 1300 km in diameter and was observed half-lit at the terminator. Its formation affected large areas surrounding the basin and also caused a tremendous amount of fracturing and surface disruption at the Caloris antipode; 180° away on the opposite hemisphere of the planet (to be discussed in detail in Section 9.6). The number of rings ranges from 3 to perhaps 6. The main ring of mountains is about 2 km high. Another faint cliff is located on the northeastern rim about 150 km from the main rim (Figure 9.9). It probably represents a fault scarp along which a block of the crust slid inward toward the excavation crater. The area between the scarp and the main rim consists of broken up material probably formed as blocks slid toward the center of the basin. Beyond the faint scarp a system of valleys radiates outward for about 1000 km (Figure 9.10). These valleys may be fault troughs or chains of large coalescing secondary craters formed from strings of basin

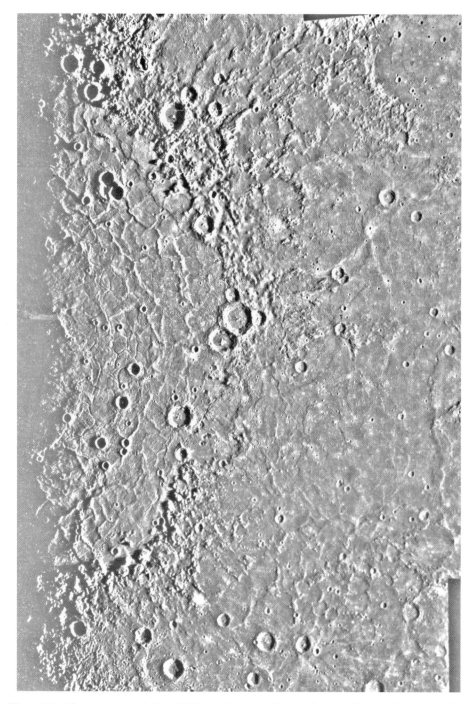

Figure 9.8. Photomosaic of the 1300 km diameter Caloris basin. This is the largest best preserved basin observed by *Mariner 10*.

Figure 9.9. This detail of the Caloris basin's northeastern rim shows the main ring located at A–A and a weaker outer ring at B–B.

ejecta. Numerous crater clusters and irregular troughs are probably secondary craters. Some of these secondaries are over 20 km in diameter.

Beyond the basin rims are several areas of hummocky plains with numerous small hills that extend outward for several hundred km. This material is probably a combination of continuous ejecta including large fragments and a significant amount of impact melt (Figure 9.11). Surrounding these ejecta deposits are smooth plains that occur up to 2500 km from Caloris. These plains are probably volcanic deposits. They are discussed in Chapter 10.

The Caloris basin floor displays a structural pattern that is unique in the Solar System. The basin interior is filled with smooth plains that are highly fractured and ridged (Figures 9.12 and 9.13). The ridges form a pattern that is both concentric and radial to the basin center. They are similar in morphology to the Moon's mare ridges. However, in the Caloris basin they are much more numerous and have a radial component not seen on the Moon. The ridges are probably caused by

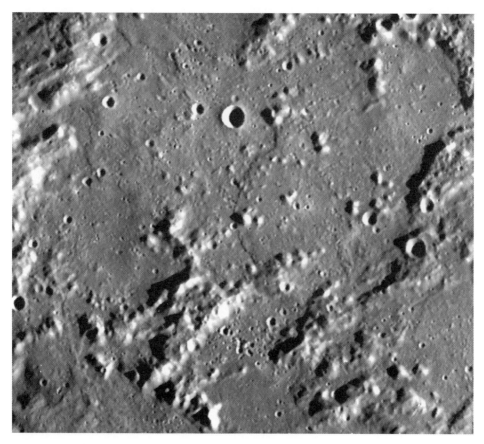

Figure 9.10. Linear valleys and ridges radiate from the rim of the Caloris basin. They were probably formed by ejecta from the impact basin.

compressive stresses as they are on the Moon. The ridges are transected by a system of younger tension fractures that also have a concentric and radial pattern (Figure 9.14). The fractures are up to 10 km wide and progressively increase in width and depth toward the basin's center (Figure 9.12). At the margin of the floor the fractures become very weak and completely disappear near the rim.

Vertical movements may account for both of these very different structures. Transection relationships indicate that the ridges formed first. They were probably caused by compressive stresses as the floor subsided. The floor covers about 30 degrees of latitude, and, therefore, has a substantial outward curvature. As a result the subsiding floor was compressed into a smaller area causing the ridges. The tensional fractures were formed next, possibly by vertical uplift that stretched the floor. Except at the Caloris antipode these tensional fractures are the only sign of tensional stresses on Mercury, making its tectonic history unique. The vertical movements may have been caused by subsidence due to the weight of crater

Figure 9.11. These rough plains with interbedded hills outside but near the Caloris basin are probably a continuous ejecta deposit of breccia and melt from the Caloris impact.

interior fill (lavas) as on the Moon. Uplift caused by upward migration of subsurface magmas may have caused the tensional stresses that formed the fractures.

9.6 HILLY AND LINEATED TERRAIN

As mentioned in Section 9.5, the Caloris basin-forming impact is also responsible for another type of terrain in a completely different part of Mercury. Directly opposite the center of the Caloris basin on the other side of the planet (the antipode) is located a peculiar, severely disrupted surface known as the hilly and lineated terrain. It covers an area seen in Mariner 10 images of at least $360,000 \, km^2$. It probably extends further. It consists of hills, depressions, and valleys that disrupt pre-existing landforms. The hills are 5 to 10 km wide and up to 2 km high. Valleys are up to 15 km wide and over 120 km long. They form a roughly orthogonal pattern trending northeast and northwest. Crater rims have been disrupted in many cases,

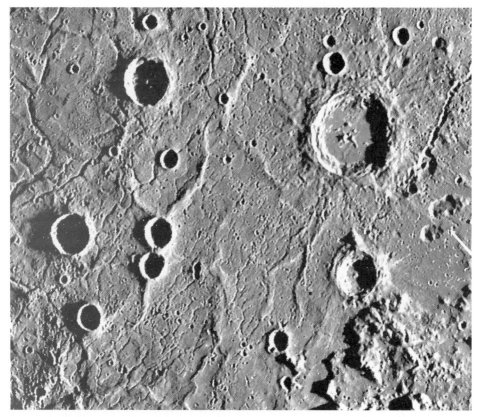

Figure 9.12. This image shows the ridged and fractured floor of the Caloris basin. The fractures transect the ridges and are therefore, younger. The irregular rimless depressions at the middle far right of the image (arrow) are probably volcanic collapse depressions. A relatively large abundance of potassium in Mercury's exosphere has been observed over the Caloris basin. It may be related to lavas from volcanic vents in turn related to the volcanic collapse depressions.

but their floors have been filled with younger plains. This indicates that volcanic activity occurred after the disrupting event (Figures 9.15, 9.16 and 9.17).

The hilly and lineated terrain is similar, but much larger in extent, to disrupted surfaces at the antipodal points of the Imbrium and Orientale basins on the Moon. The fact that these terrains occur at the antipodal points of large impact basins strongly suggests that they are the result of the impacts. Seismic waves generated by these impacts converge or focus at the antipodal regions (see Figure 9.18). Computer simulations of seismic wave propagation for impacts of this size show that the seismic effects in the antipodal regions can be enormous. The ground may experience vertical motions greater than 1 km in a matter of minutes, and tension fractures rend the crust to depths of tens of kilometers. This stress breaks the surface into a jumble of blocks and depressions like the hilly and lineated terrain. Models

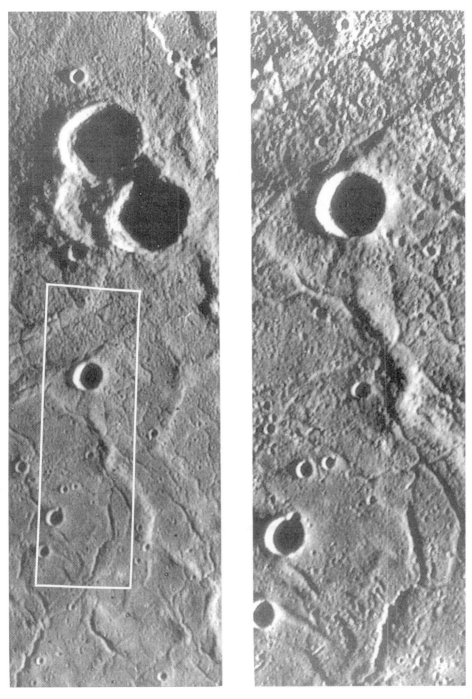

Figure 9.13. These third-encounter images show a small portion of the Caloris basin floor. The rectangle in the image on the left indicates the location of the image on the right.

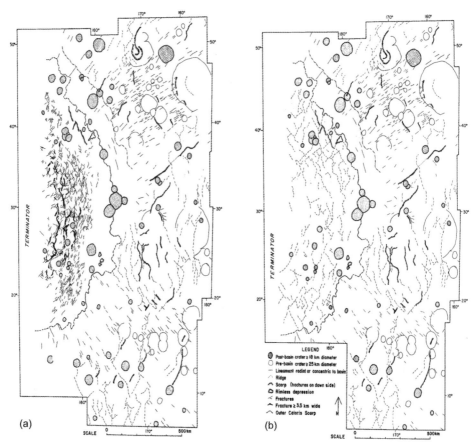

Figure 9.14. These maps show the pattern of fractures (a) and ridges (b) on the Caloris basin floor. Both systems show a radial and concentric component (from Strom, Trask, and Guest, 1975).

show the effects are enhanced by seismic waves refracted by Mercury's enormous iron core. This explains why the hilly and lineated terrain is much more extensive on Mercury than the Moon. Furthermore, fractures penetrating to great depth could provide egress to the surface for lavas that appear to have flooded the low lying areas within craters after their disruption. As at Caloris, enhanced potassium in Mercury's exosphere has been observed over the antipodal hilly and lineated terrain. It may be coming from the rocks formed from the lavas or from tension fractures.

9.7   ORIGIN OF IMPACTING BODIES

9.7.1   Asteroids

The origin of the objects responsible for the cratering record in the inner Solar System is somewhat controversial. Today the only objects that cross the inner

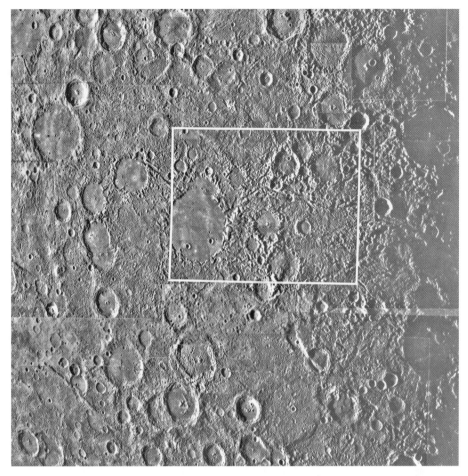

Figure 9.15. This photomosaic shows the region of Mercury's hilly and lineated terrain. The outlined area is the location of the higher resolution image shown in Figure 9.16.

planet orbits are comets and high eccentricity asteroids called Amors and Apollos. There is no doubt that they have, and still are, contributing to the cratering record. But are they the only source?

9.7.2   Elusive vulcanoids

Some people have speculated that there is a population of vulcanoids, or rocky bodies orbiting around the Sun closer to the Sun than Mercury. Perturbations on the Vulcanoids could cause impacts with the surface of Mercury. Recent searches have not found any, and it is likely that they do not exist.

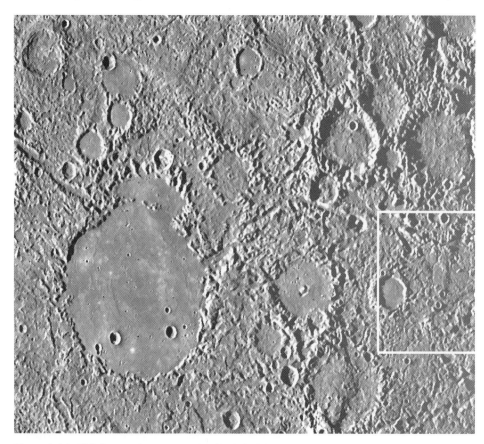

Figure 9.16. This image shows more detail of the hilly and lineated terrain. The smooth plains filling the large crater at left are younger than the hilly and lineated terrain. The outlined area can be seen at high-resolution in Figure 9.17.

9.7.3   Evidence for two collisional populations

Comparisons of the Solar System cratering record and dating of returned lunar rocks, including lunar meteorites, have provided some information on the origin of impactors. The manned *Apollo* missions to the Moon returned rocks from a variety of locations. From these samples it was learned that the relatively sparsely cratered mare lavas date from about 3.9 to 3.0 billion years old. The heavily cratered highlands are even older, dating from about 4.4 to 4.0 billion years. The lunar highlands accumulated their great abundance of craters, including the large mare-filled basins, over a geologically short time span of no more than 400 million years. On the other hand, the younger lunar maria accumulated their much smaller number of craters over the enormous span of 3 to 4 billion years (about 10 times longer than the luner highlands). This must mean that the Moon experienced a period of intense bombardment that ended early in its history about 3.9 billion years ago. It was

Figure 9.17. This is one of the highest resolution images of the hilly and lineated terrain taken by *Mariner 10*. It shows a broken-up surface of hills and valleys. The hills range from 0.1 to 1.8 km high. The large crater on the left is 31 km in diameter.

during this intense period of bombardment that most basins were formed. Since that period ended, large impacts have been relatively infrequent and no large basin-forming events have occurred.

There is some evidence that the period of heavy bombardment was a catastrophic event rather than a rapidly declining high flux of objects. Impact melts from 3 to possibly 6 impact basins indicate they were formed between 3.88 and 4.05 billion years ago. Furthermore, additional analyses of *Apollo* samples indicate the U-Pb and Rb-Sr systems were disturbed ~3.9 billion years ago. Recently analysed lunar meteorites also have impact melt that dates from about 3.9 billions years ago. These data suggest there was a catastrophic bombardment about 3.9 billion years ago that not only affected the Moon, but almost surely affected all the inner planets, including Mercury. Analyses of lunar impact melts indicate that at least one of these projectiles had a differentiated iron-rich core. Meteorite analyses

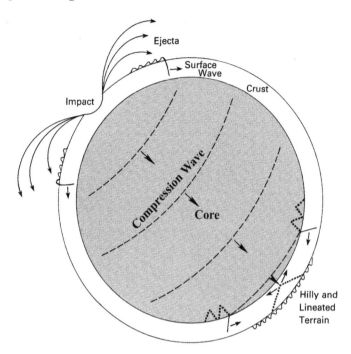

Figure 9.18. This diagram shows the probable cause of the hilly and lineated terrain. Seismic waves generated by the Caloris impact were focused at the antipodal point, causing large vertical ground movements resulting in the hilly and lineated terrain (courtesy Peter Schultz, Brown University).

indicate that the asteroids were also heavily cratered about 3.9 billion years ago. These data suggest the origin of the objects was the asteroid belt.

The impact crater size distributions for Mercury and other inner Solar System objects seem to be consistent with an early cataclysmic heavy bombardment. Crater size/frequency distributions measure the number of craters within certain size ranges. This in turn is a measure of the size distribution of the impacting objects when proper scaling relationships are taken into consideration. Crater abundances derived from size/frequency distributions are also used to date surfaces relative to each other, and also on an absolute time scale if the crater production rate is known.

The crater size/frequency distribution is conveniently displayed on what is called a "Relative" (R) plot. This type of plot was devised to better show the size distribution of craters, and the crater number densities for determining relative ages. On an R plot the size/frequency distribution is normalized to a differential −3 distribution function, or slope. The reason a −3 reference distribution is used is because most impact crater size/frequency distributions are within ±1 of a −3 distribution. On an R plot a differential −3 distribution plots as a horizontal straight line. The vertical position of the line is a measure of the crater density or relative age; the higher the

Table 9.1. Size/frequency distributions for slopes of −2, −3, and −4.

Crater diameter (km)	Slope		
	−2	−3	−4
64	1	1	1
32	4	8	16
16	16	64	256
8	64	512	4096
4	256	4096	65536

vertical position, the higher the crater density and the older the surface. On an R plot a line sloping to the left at an angle of 45° is a differential −2 distribution, and one sloping to the right at 45° is a differential −4 distribution. Usually the data are binned into $\sqrt{2}$ increments because there are many more craters at small diameters than large diameters. For example, a distribution with slope −3 would have 1 crater of diameter 64 km, $(1/2)^{-3}$ craters (8) of 32 km diameter, $8 \times (1/2)^{-3}$ craters (64) of 16 km diameter, and $64 \times (1/2)^{-3}$ craters (512) of 8 km and so on. This is displayed in Table 9.1 along with size/frequency distributions for slopes of −2 and −4. Figure 9.19 is a diagramatic representation of the difference between a −3 and a −2 slope.

Mathematically, the R value is expressed as follows:

$$R = \frac{D^3 N}{A(b_u - b_l)},$$

where D is the geometric mean diameter of the size bin, N is the number of craters in the size bin, A is the area counted, b_u is the upper limit of the size bin, and b_l is the lower limit.

The heavily cratered surfaces of the Moon, Mars, and Mercury represent the period of heavy bombardment early Solar System history. These surfaces on Mercury, the Moon, and Mars all have similar crater distributions (Figure 9.20). They show a complex curve with about a −2 distribution at diameters less than about 50 km, a −3 distribution between 50 and 100 km, and about a −4 distribution between 100 and 500 km. At diameters greater than 500 km the statistics are too poor to determine a crater distribution with any confidence. One notable difference between the curve for the Moon and those for Mercury and Mars is that at diameters less than about 50 km there is a marked deficit of craters on Mercury and Mars compared to the Moon. This is almost surely due to the emplacement of intercrater plains material on Mercury and Mars compared to the Moon where these plains are extremely rare (see Figure 9.21). This strongly suggests that inter-crater plains formation on Mercury was occurring during the period of heavy bombardment. The youngest smooth plains surfaces that surround and fill the Caloris basin also show a similar crater size/frequency distribution as the

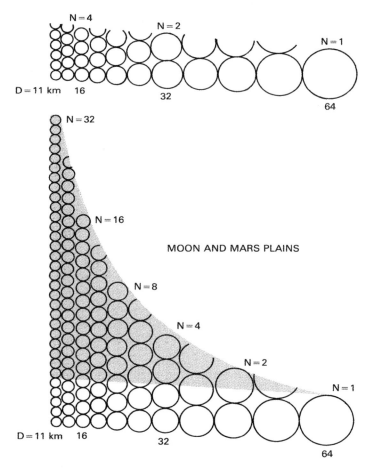

Figure 9.19. This diagram illustrates the difference between the size/frequency distributions of craters between 11 and 64 km in diameter found on the terrestrial planets. The size distribution in the upper diagram represents the heavily cratered highlands of the Moon that resulted from the period of heavy bombardment. It has a differential −2 slope. Somewhat similar distributions occur in the heavily cratered terrain on Mars and Mercury but the slopes are more like a −1.5 differential slope. The size distribution in the lower diagram represents the lightly cratered plains on the Moon and Mars. It has a differential −3 slope. The shaded area indicates the difference between the two crater populations. (See text for explanation.)

highland, but at a lower density (Figure 9.20). The crater density on these younger surfaces is much greater than on the lunar maria. The post-Caloris curve is similar to that of Mercury's highlands but it is shallower because it has not been effected by plains emplacement. It is at a lower level because the post Caloris surface is younger than the highlands, and its shape indicates that it is part of the period of heavy bomardment (probably near the end of that period).

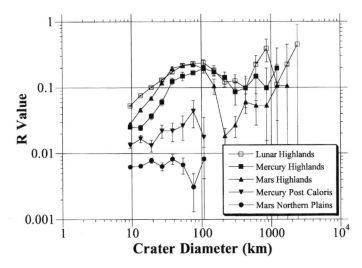

Figure 9.20. This R plot is a comparison of the crater size/frequency distribution of the lunar, Mercurian, and Martian heavily cratered highlands. They all have a similar shape indicating a common origin. The steeper slopes for Mercury and Mars at smaller diameters are the result of obliteration of craters by intercrater plains formation. Also shown is the size distribution of the post-Caloris crater population and the lightly cratered relatively young surfaces on Mars (see text for explanation).

   A comparison of the *R* plots of the highland cratering records on Mercury, the Moon and Mars describes the nature of the objects that were impacting during the period of heavy bombardment within the inner solar system. The curves all have similar shapes except at diameters less than about 40 km where intercrater plains emplacement has modified the curves for Mercury and Mars as mentioned above. However, at diameters between about 40 km and 150 km, where the curves are probably unaffected by plains emplacement, the curves are laterally displaced with respect to each other. In fact, they are displaced in a manner that requires high velocities for planets at smaller heliocentric distances; larger craters on Mercury and smaller craters on Mars compared to a given size crater on the Moon. The best fit of the curves shows that for a 100 km diameter crater on the Moon, the crater size is 120 km diameter on Mercury and 80 km diameter on Mars. A comparison of the ratios of impact velocities derived from scaling laws and their required eccentricities suggest that the objects responsible for the period of heavy bombardment were confined to the inner solar system with semimajor axes between about 0.8 and 1.2 AU (Figures 9.22 and 9.23).

9.7.4   Surfaces younger than the period of heavy bombardment

Mercury may have some surfaces younger than the period of heavy bombardment on the unimaged portion of the planet. We will have to await further exploration to answer that question.

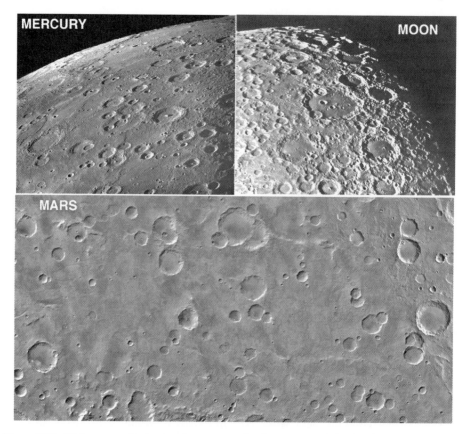

Figure 9.21. This composite image shows extensive intercrater plains in the heavily cratered highlands of Mercury (upper left) and Mars (bottom), but little or no intercrater plains on the Moon (upper right). The implacement of intercrater plains on Mars and Mercury probably resulted in the greater paucity of craters at diameters less then about 40 km compared to the Moon. The individual images are not to scale.

Young surfaces on the Moon and Mars have a significantly different size/frequency distribution. They show a −3 distribution in the diameter range of about 1 to 100 km. There are very few craters larger than 100 km on these young surfaces (see Figure 9.20). At least on Mars, and probably on the Moon, this population of craters is most likely the result of impacts from the *collisionally evolved* asteroid belt.

Since the objects responsible for the period of heavy bombardment have a different size/frequency distribution, they appear to come from a different population, but one confined to the inner Solar System. One possibility is they were primordial, *collisionally unevolved* asteroids that were dynamically ejected from the asteroid belt by the combined gravitational perturbations of Jupiter and planetary embryos retained from the formation of the inner planets. Another possibility is that

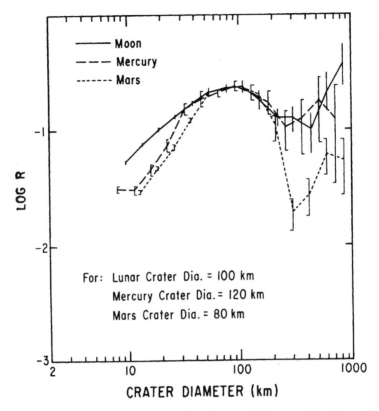

Figure 9.22. The crater size/frequency distributions for the highlands of the Moon, Mars, and Mercury (Figure 9.20) have been matched from about 40 km to 150 km diameter (the range not effected by intercrater plains emplacement and having good statistics). The lateral shifts in the curves require higher planet impact velocities with decreasing heliocentric distance; larger craters on Mercury and smaller ones on Mars compared to a given size crater on the Moon (from Strom and Neukum, 1988).

they could be fragments from a giant collision in the asteroid belt very early in its history. Either of these origins could provide a cataclysmic bombardment of the inner planets. However, there may by other ways to produce this ancient population of objects.

9.8   RELATIVE AND ABSOLUTE AGES

The crater abundance superposed on various geologic units can be used to determine the age of a surface relative to other geologic units. This technique, together with embayment relationships among units and transection relationships between tectonic structures and various units, forms the basis for determining the order of

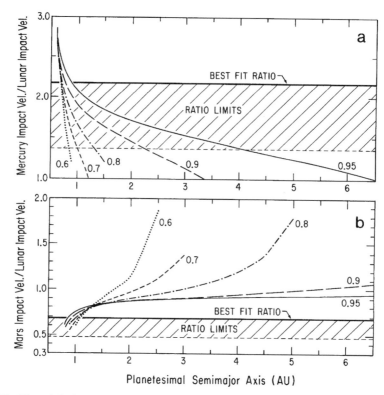

Figure 9.23. Plot of the impact velocity ratios Mercury/Moon (a) and Mars/Moon (b) derived from matching the highlands crater curves (Figure 9.22), verses impactor semimajor axes for eccentricities from 0.6 to 0.95. The hatched areas are the limiting impact velocity ratios for an acceptable curve fit, while the solid horizontal lines are the ratios derived from the best curve fit shown in Figure 9.22. Only planetesimals with semimajor axes between about 0.8 and 1.2 AU lie within the same region of the impact velocity ratio limits. Jupiter crossers (objects that cross the orbit of Jupiter) have semimajor axes greater than 2.7 AU (from Strom and Neukum, 1988).

emplacement (the relative age) of geologic units. The geologic maps of Mercury are based on these techniques. The age of a surface based on the cratering record requires that the surface is: (1) not saturated with craters (i.e., it is a *production crater population*), (2) that only superposed craters are counted (no relic or ghost craters from an underling unit); and (3) that all secondary and volcanic craters are eliminated from the counts.

   If one knows the rate at which craters are formed, then the age of the surface can be determined. The rate of crater formation depends on a knowledge of the rate at which various objects collide with the planet or satellite. This, in turn, depends on the origin of the impacting objects, and the proportion of each type of impactor (e.g., comet or asteroid, that has impacted the planet). Obviously, estimates of these factors contain large uncertainties and, therefore, the estimated *absolute ages* are uncertain. However, on the Moon, where surfaces have been dated from returned samples, it has been possible to date surfaces by comparing the crater abundances on surfaces of known absolute ages to derive a crater production function that can be used to measure the ages of other surfaces where no rocks have been returned. This works quite well on the Moon, but extrapolating this crater production rate to other planets can result in significant errors. One must first assume that the impacting objects were the same at the Moon and the planet, and then scale the production function by certain scaling laws. However, we know that the terrestrial planets have been impacted by at least two populations of objects, comets and asteroids. If the period of heavy bombardment was the result of a catastrophic event then a third population may be involved. Furthermore, there are two crater populations in the inner solar system, one for younger surfaces and another for ancient surfaces. Any extrapolations must use the correct crater population at both bodies. In the outer Solar System the problem is even more complex and extremely uncertain.

9.8.1 Mercury's surface is ancient

On Mercury, absolute ages are derived from those determined for the Moon. The dependence of the Mercurian cratering rate is assumed to be the same as for the Moon. Also considerations of asteroid and comet impact probabilities at Mercury, and corrections for impact velocities, scaling, and gravitational focusing effects are taken into account. It is obvious that there can be relatively large errors in the age determinations. However, it is not as bad as it seems. The period of heavy bombardment almost surely ended on Mercury at the same time it ended on the Moon; before about 3.8 billion years ago. Therefore, surfaces that show a crater population associated with the period of heavy bombardment must be $\geq$3.8 billion years. Since all surfaces on Mercury explored to date show this crater population, they are probably between 3.8 and 4.5 billion years old. Surface ages derived for units with different crater densities are extrapolated between these extremes with the Caloris basin assumed to have formed 3.8 billion years ago.

9.8.2 Will there be younger terrains on the unimaged side?

One must be very cautious because we have only seen 45% of the surface and only about 25% of the surface was viewed at sun angles suitable for terrain analysis. Other areas of Mercury could be considerably younger than those where crater counts are currently available.

10

Plains: smooth and intercrater

10.1 WILL THE UNSEEN SURFACE OF MERCURY BE COVERED WITH PLAINS?

As soon as the first pictures of Mercury were available from Mariner 10, it was realized that the surface was largely covered with plains. The plains were obviously heavily cratered and from this it was inferred that they were emplaced early in Mercury's history. About 60% of the imaged portion of Mercury is plains, and if this proves true for the unexplored side, then plains are the most abundant terrain on Mercury. The other 40% consists of large craters, basin rim structures, and the hilly and lineated terrain. Mercury's plains differ from the Moon's plains in two important respects (recall that the lunar maria are plains consisting of basaltic lava flows). First of all, Mercurian plains are much more widespread than at the Moon, and second, they appear to have a higher albedo than those on the Moon.

10.2 MERCURY HAS TWO TYPES OF PLAINS

The Mercurian plains are usually divided into two general types:

(1) intercrater plains; and
(2) smooth plains. This division is largely based on differences in crater abundances and mode of occurence between the two types. Intercrater plains are much more heavily cratered than smooth plains, and are, therefore, older. They occur in the Mercurian highlands between clusters of craters (Figures 10.1, 10.2 and 10.3). Like the lunar maria, smooth plains are relatively young and largely confined to the interior and/or exterior of impact basins and large craters (Figure 10.4).

Plains can have various origins. They can be caused by erosion and deposition similar to Earth's plains. They can also be formed by impact ejecta mantling such

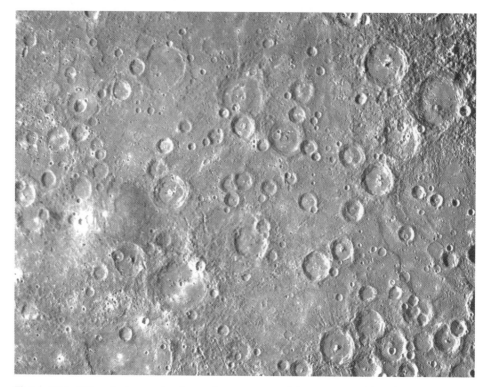

Figure 10.1. This photomosaic shows large areas of intercrater plains between heavily cratered areas. Intercrater plains dominate Mercury's highlands.

as some highland plains on the Moon. Most plains are the result of volcanism that has produced enormous flood basalt deposits, such as those on the Moon, Venus, Earth, and Mars. The origin of Mercury's plains is still uncertain, but current evidence seems to favor a volcanic origin for most.

10.3   THE INTERCRATER PLAINS, SOME DETAILS

10.3.1   Distribution, age, and morphology

About 45% of the surface viewed by *Mariner 10* is occupied by intercrater plains. They occur between and around clusters of large craters in the heavily cratered highlands of Mercury.

In many cases the superposed small craters occur in chains or clusters suggesting they are secondary craters from the larger craters and basins that make up the highlands. The high crater density indicates that these plains predate the smooth plains and that they constitute one of the oldest surfaces on the planet. They apparently span a range of ages contemporaneous with the period of heavy

Figure 10.2. This image of the Mercurian highlands shows large areas of intercrater plains. The large scarp near the planet's limb is a thrust fault.

bombardment. They partially bury some large craters (Figure 10.3) and the ejecta blankets of others, but other large craters and their ejecta are superposed on the plains. The older and intermediate age craters seem to be more affected by intercrater plains formation than the younger fresh craters which are the ones that contribute most to the superposed secondary cratering. Figure 10.4 (colour section) shows paleogeologic maps by Martha Leake of Mercury's incoming side as viewed by Mariner 10. They show that intercrater plains were emplaced during the formation of older class 5 through class 3 craters and that the volume of plains generally decreases as age decreases. Secondary craters superposed on the intercrater plains appear to be derived mainly from the younger class 1–3 craters and class 4 basins of the heavily cratered terrain. Patches of plains emplaced during the formation of the young class 1 and 2 craters are equivalent to smooth plains. These relationships strongly suggest the intercrater plains were emplaced during the period of heavy bombardment, and that the volume of these plains decreased as the impact rate declined. In fact, the formation of intercrater plains and smooth plains may not represent two distinct episodes of plains formation. Instead they may represent a more-or-less continuous period of plains formation lasting from sometime during the period of heavy bombardment to the emplacement of the youngest smooth plains. Absolute dating of intercrater plains based on their crater densities suggests that they formed between 4.0 and 4.2 billion years ago (see Chapter 9), but they may have continued as smooth plains to at least 3.8 billion years ago.

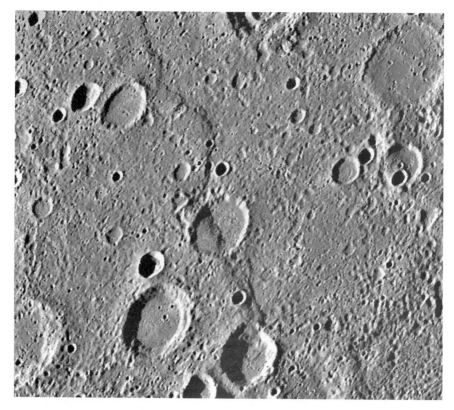

Figure 10.3. A high-resolution image of the intercrater plains. It shows an abundance of small craters that are mostly secondary impact craters from younger primary impact craters. Also shown are intercrater plains that have flooded a large crater (70 km) in the upper right-hand corner of the image.

10.3.2   Crater degradation by intercrater plains

The crater/size frequencey distribution indicates that a substantial fraction of the craters less than 50 km in diameter have been destroyed by intercrater plains formation. There is a significantly smaller abundance of craters less than 50 km diameter on the Mercury highlands compared to the lunar highlands, showing a systematic loss of craters relative to the Moon, where there are very few intercrater plains (see Chapter 9). This is exactly what would be expected if Mercury's inter-crater plains were formed during the period of heavy bombardment with smaller craters buried first, followed by larger craters as the plains continued to be deposited (Figures 10.3 and 10.4, colour section). Intercrater plains form level, to gently rolling surfaces with a high density of craters greater than about 15 km diameter. This high crater denisty gives these plains a rough appearance and attests to their great age.

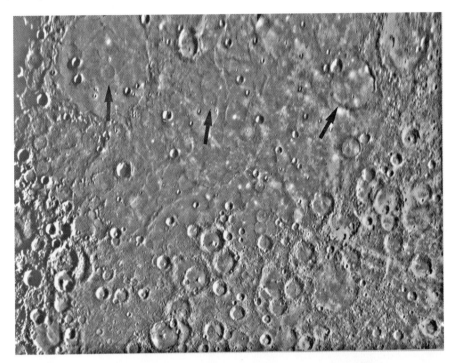

Figure 10.5. This photomosaic shows a large area of smooth plains that occurs in the north polar region and appears to fill the ancient Borealis basin about 1500 km in diameter. The plains also fill the younger Goethe basin (340 km diameter) located in the upper left corner of the photomosaic. Some craters that have been flooded by smooth plains are indicated by arrows.

10.4   SMOOTH PLAINS, SOME DETAILS

10.4.1   Distribution, age, and morphology

Smooth plains cover almost 15% of the imaged portion of Mercury. About 90% of the smooth plains are associated with older large impact basins, but they also fill smaller basins and large craters (Figures 10.5, 10.6, 10.7 and 10.8). There are two large concentrations of smooth plains on the explored side of Mercury. One fills and surrounds the 1300 km diameter Caloris basin. The other occupies a large highly degraded impact basin about 1500 km diameter called the Borealis basin (Figure 10.5). Smooth plains have the lowest crater density, and are, therefore, the youngest geologic unit on Mercury.

The smooth plains on Mercury appear similar in morphology and mode of occurrence to the lunar maria. Craters within the Borealis, Tolstoj, and other basins have been flooded by smooth plains. Furthermore, the crater densities on the floors of these basins are significantly less than those on the terrains immediately surrounding the basins (Figures 10.5 and 10.8). These observations alone indicate

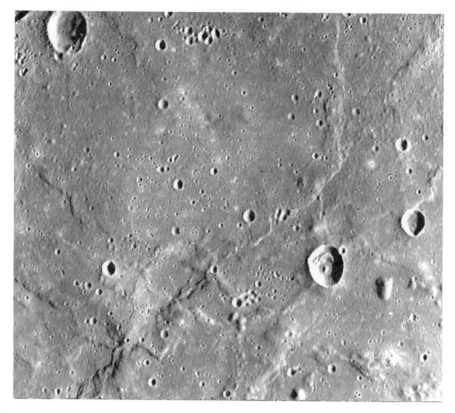

Figure 10.6. Most of the area in the image is covered by smooth plains that surround the Caloris basin. A ridge similar to a lunar wrinkle ridge crosses the middle of the image. Wrinkle ridges are compressional tectonic features.

that smooth plains are younger than the basins they occupy and cannot be impact melt that resulted from the impacts that formed the basins. Smooth plains filling smaller craters and basins could be impact melt, however.

Based on the shape of the crater size/frequency distribution and the fairly high crater density (Figure 9.20), the smooth plains probably formed near the end of heavy bombardment. They may have an average age of about 3.8 billion years, or several hundred million years younger than the intercrater plains laid down during the period of heavy bombardment. This means they are, on average, older than the flood basalt lavas that consitute the lunar maria.

10.5 ORIGINS OF PLAINS

Two radically different origins of the Mercurian plains have been proposed. One hypothesis considers both the smooth and intercrater plains to be impact ejecta deposits from large basins, similar to the Cayley plains on the Moon. The other

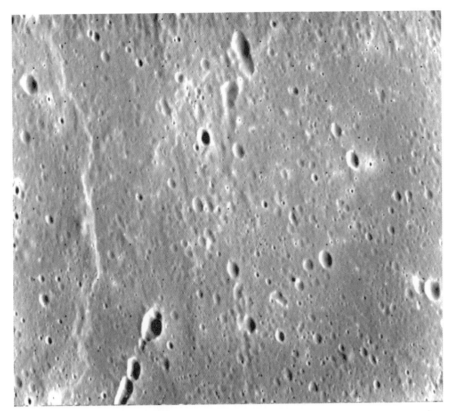

Figure 10.7. High-resolution image of the smooth plains. The smallest crater that can be seen is about 200 m across. These plains surround the Caloris basin.

hypothesis considers both plains units to be volcanic deposits. In the past, the origin of plains on Mercury has been controversial, but a re-evaluation of the *Mariner 10* color data together with earlier stratigraphic analyses favor a volcanic origin, particularly for the smooth plains.

The problem with determining the origin of the plains is the dearth and relatively low quality of the available data. The mode of formation of any plains unit is difficult to determine because there are usually not many landforms that are diagnostic of their origin. For instance, the origin of plains on the Moon was very controversial when only relatively low-quality earth-based telescopic observations were availible. That is about where we stand for Mercury at the present time. However, the great advantage now is that we can draw on our past experiences with the Moon.

10.5.1 Mercury's smooth plains as impact basin ejecta deposits or impact melt

On the Moon just beyond the continuous ejecta blankets of fresh impact basins lie small areas of light smooth plains. These deposits are fluidized basin ejecta that flowed beyond the continuous ejecta blanket and buried some pre-existing craters. At greater distances occur patches of gently rolling light plains that fill craters and other low areas in the lunar highlands. Except for their small areal extent, they resemble to some degree the intercrater plains of Mercury. *Apollo 16* returned samples from one of these areas that proved to be impact breccias, probably from the Imbrium basin.

Under this hypothesis, the smooth plains surrounding the Caloris basin are smooth ejecta deposits from the basin. The interior smooth plains filling the Caloris basin would be impact melt under this hypothesis. Other smooth plains filling older basins would be either impact melt or ejecta deposits from other basins. Furthermore, the intercrater plains would be the same type of ejecta deposits as the Cayley plains on the Moon and would have been formed early in Mercury's history during the period of heavy bombardment. The similar albedos of smooth and intercrater plains has been used as an arguement in favor of an impact origin. This hypothesis for plains formation implies there has been no volcanism on Mercury's explored side.

10.5.2   Mercury's smooth plains as volcanic deposits

There are several severe problems with the impact ejecta/melt hypothesis. Lunar light plains ejecta deposits only cover about 5% of the entire Moon's surface. In contrast, Mercurian smooth plains cover about 15% of the imaged side, which may be representative of the planet as a whole. It is difficult to explain why Mercury's smooth plains, if ejecta deposits, are more widely distributed than on the Moon when the ballistic range is much less. For example, the smooth plains surronding the Caloris basin extend outward from the rim in a more-or-less continuous deposit for more than 2000 km in some areas. No such extensive ejecta deposits occur around lunar craters of comparable size. Since the ballistic range on Mercury is about half that for the Moon, the ejecta blankets should be closer to the rim on Mercury, as observed for smaller craters. Furthermore, the surface gravity on Mercury is 2.3 times stronger than the Moon's so flow of ejecta deposits beyond the continuous ejecta blanket should be much more restricted on Mercury for comparable impact melt viscosities. Also, large fresh craters greater than 200 km diameter do not show a proportionately larger amount of smooth plains surrounding them. They, instead, show ejecta deposits consisting of narrow, continuous deposits with numerous clusters and strings of secondaries right up the the crater rims.

There are several aspects of the large expanses of smooth plains that indicate they are volcanic in origin. One observation that is very damaging to the ejecta hyposthesis is the observation that the smooth plains are younger than the basins they occupy or surround. Ejecta deposits and impact melt are part of the impact process, and, therefore, are the same age as the basin to which they are associated. In the interiors of the large Tolstoj and Borealis, basins, smooth plains have embayed older craters on their floors indicating that the plains are younger than the basins they occupy (Figures 10.5 and 10.8). Crater counts on the smooth plains

Figure 10.8. Smooth plains fill the ancient Tolstoj impact basin 400 km in diameter. The arrows point to craters that have been flooded by the smooth plains indicating they are younger than the basin. The "A" indicates the location of a rimless, elongate depression that may be volcanic.

surrounding the Caloris basin and on Caloris ejecta deposits show they are of different ages, therefore, these smooth plains cannot be ejecta deposits from Caloris. Furthermore, there appears to be a slight difference in albedo between the Caloris smooth plains and the intercrater plains, with the smooth plains being slightly darker. Earth-based radar observations show that the annulus of smooth plains surrounding Caloris is strongly down-bowed like the lunar maria. On the Moon this is due to the weight of the volcanic iron-rich basalts. Also, several irregular rimless depressions on the floors of the Caloris and Tolstoj basins appear to have a volcanic origin (Figure 10.8 and Figure 9.12). *Mariner 10* enhanced color images show that the boundary of smooth plains within the Tolstoj basin is also a color boundary suggesting a different composition from the surrounding material. However, color differences could indicate age difference due to varying amounts of space weathering or even textural differences such as grain size or porosity. The youth of smooth plains relative the the basins they occupy, their great areal extent, the presence of a sharp color boundary in at least one case, and other stratigraphic relationships indicate that the majority of smooth plains are volcanic deposits which erupted about 3.8 billion years ago.

Recent ground-based imaging has captured Mercury images at moments of excellent atmospheric stability showing albedo differences between regions on the

Figure 10.9. Good contrast between bright and dark regions is in this ground-based image of Mercury. The image was captured by Frank Melillo with his Celestron 8 inch Starlight Express and commercially available CCD imaging camera from his backyard in Holtsville, NY. Tim Wilson, another amateur Mercury observer prepared the latitude and longitude grid for the image for better identification of surface features. The grid lines are separated by 22.5° in latitude and 30° in longitude, with the central meridian at 158° longitude. The dark regions coincide with the smooth plains, above, and to the west of Tolstoj.

surface (see Figure 1.7 and 1.8). Amateur observers are also taking advantage of imaging Mercury with relatively inexpensive and very capable instrumentation. Some images show definite albedo differences across the surface, some associated with bright impact craters, also associated with radar bright features. Darker regions may be examples of smooth plains. A ground-based image of Mercury is shown in Figure 10.9. The image was obtained with a back-yard Celestron telescope and imaging CCD from Holtsville, NY. A grid has been overlain to permit knowledge of the proper latitudes and longitudes on the planet and permit identification of surface features. The vertical line in the center of the image is the 158° longitude. The large darker albedo region coincides with a large expanse of smooth plains replete with wrinkle ridges and embayed craters. The region encompasses smooth plains surrounding the Caloris basin. It includes Tir Planitia on the west, Odin Planitia, spanning the meridian, and Budh Planitia to the east. The large, smooth plains in the interior of Tolstoj are in the lower portion of the dark region just west of the central meridian. This observation is corroborated by *Mariner 10* image photometric studies of the smooth plains surrounding the Caloris basin. They indicate that these smooth plains are slightly darker than the adjacent intercrater plains.

Definitive volcanic landforms are rare on Mercury, but one must keep in mind that *Mariner 10* resolution and coverage is only about the same as Earth-based telescopic resolution and coverage of the Moon before the advent of spacecraft exploration. At this resolution, there are few identifiable volcanic landforms on the Moon although most of the frontside is covered by volcanic deposits. There are, however, some irregularly shaped rimless pits on smooth plains that have been interpreted to be volcanic. They are similar to volcanic rimless pits on the Moon and shown in Figure 10.8 and 9.12.

These combined observations strongly suggest that most of the smooth plains have a volcanic origin. The distribution of the smooth plains is, in fact, very similar to the lunar volcanic maria; both are associated with large impact basins and have about the same areal distribution. This further indicates that they were emplaced in a similar fashion.

10.5.3   Mercury's intercrater plains as volcanic deposits

The origin of intercrater plains is considerably more difficult to interpret. Similar arguments can be made for the volcanic origin of Mercurian intercrater plains as are made for smooth plains. They cover about 45% of the imaged side of Mercury, but there is no evident source basins from which impact deposits could be derived. On the Moon small patches of intercrater-like plains occur in the highlands but they only constitute about 3% of the Moon's entire surface. The origin of these lunar deposits is still debatable but, unlike the Mercurian deposits, they at least can be indentified with possible source basins.

As discussed in Chapter 9, the frequency of the interior morphologies of craters on the smooth plains, the lunar maria, and the intercrater plains are the same, but differ signifantly from that of the lunar highlands. This suggests that the intercrater plains have physical properties (strength and cohesiveness) similar to the lunar maria volcanic deposits and probably those of the volcanic smooth plains (see Figure 9.3). In other words, they are much more similar to coherent igneous rock than the lunar highlands megabreccia.

Newly recalibrated *Mariner 10* images bear strongly on the origin of Mercury's plains deposits. The recalibrated image mosaics show the complete ultraviolet-orange color data for the region near Kuiper crater. The data are interpreted in terms of visible color reflectance for iron-bearing silicate regoliths. Based on the albedo and color ratios (UV/orange), the content of opaque minerals and soil maturity were estimated. Many differences across the surface are dramatically displayed in Figure 10.10 (colour section). Smooth plains units near the Rudaki crater show distinct color boundary embayment relationships that correspond to previously mapped plains boundaries. This strongly indicates that these plains have a different composition, age, or grain size and porosity than the surroundings and are probably deposited as volcanic flows (Figure 10.10). It is not possible to determine from the ratioed images whether the plains are basaltic, but the morphology suggests that they were formed from a relatively fluid lava, one example of which is basalt. This area also shows two areas of spectrally blue material with high opaque indecies, low albedos and diffuse boundaries. Since there are no impact craters associated with this material they could be more mafic pyroclastic deposits. Kuiper and Muraski craters show a very low opaque mineral index which may indicate they have excavated into an anorthositic crust. These observations strongly indicate Mercury is compositionally heterogeneous and that volcanism has contributed to plains formation. This is strengthened by Earth-based spectro-graphic data which was discussed in detail in Chapter 8.

10.6 MODES OF VOLCANIC PLAINS FORMATION

10.6.1 Earth, Mars, and the Moon

The form that volcanism takes is highly dependent on the composition of the lava, and the volume and rate at which it is erupted. The more silica a magma contains in proportion to its *bases* (iron (Fe), magnesium (Mg), calcium (Ca), sodium (Na), etc.) the higher its viscosity. Silica forms tetrahedra which are bound together in the liquid. This tetrahedral network resists the change of shape of the liquid, and, therefore, restricts its freedom to flow. With a large content of bases, chaining of tetrahedra is less extensive and flow is easier.

On Earth, lavas rich in Si, aluminum (Al), Na, and potassium (K) and poor in Fe and Mg are extremely viscous and form short, thick flows or domes. They usually contain large amounts of gas (primarily water vapor and carbon dioxide [CO_2]) that can be released explosively to form extensive ash deposits. This type of volcanism can form two major volcanic landforms, composite volcanoes (also called strato-volcanoes) and ash flow plains. Composite volcanoes are composed of alternating layers of lava (usually andesite) and ash that form large conical mountains like Fujiyama in Japan. Ash flows are enormous volumes of ash that are erupted from the flanks of composite volcanoes resulting in their collapse and formation of calderas.

Lavas rich in Fe and Mg and with SiO_2 content varying from about 45–57% are very fluid. These basaltic lavas are composed of plagioclase feldspar, olivine and pyroxene. But each of those three constituents comes in multiple compositions and crystalline structures. Thus one basalt may be different chemically and mineralogi-cally from another even though they share a similar broad mineralogy of pyroxene, plagioclase feldspar, and olivine. Basalt eruptions can produce either huge domical mountains with summit calderas, called shield volcanoes, or extensive plains called plateau or flood basalts. On Earth, shield volcanoes are formed over mantle plumes whose positions are stable beyond the timescale of plate tectonics. Therefore, plates move over these plumes to produce chains of shield volcanoes. The Hawaiian Islands with its huge shield volcano Mauna Loa is the best known example. Large shield volcanoes also occur on Mars and Venus, but none occur on the Moon and none have been observed on Mercury.

Flood basalts on Earth are also thought to be associated with unusually large mantle plumes called superplumes. They are the result of voluminous fissure eruptions of basalt that produce extensive plains covering thousands of square kilo-meters. The fissures that give rise to these flows are usually covered by the lava so the sources of the flows are rarely visible. Large deposits of flood basalts occur in India, Brazil, Africa, and the United States. In the United States, the Columbia River flood basalts cover about 220,000 km^2 of Washington, Oregon, and Idaho. Individual flows average about 25 m thick, and the estimated volume of these deposits totals

$195,000\,km^3$.They were erupted about 15 million years ago. Flood basalt eruptions are the most common type of volcanism on Mars, Venus, and the Moon. There, they produce vast areas of thick basaltic lava plains. The lunar maria, the plains of Venus, and the intercrater plains and much of the Northern Plains of Mars are probably flood basalt deposits. Martian meteorites all have a basaltic composition, as do the returned *Apollo* mare samples. The lunar maria are flood basalts that were erupted over a period of more than one billion years. They cover about 17% of the surface with an estimated volume of 10 million km^3. Some individual lava flows have traveled more than 350 km. Venus flood basalts cover about 70% of the planet.

10.6.2   Terrestrial magmas and temperature

The way in which the many forms and composition of minerals combine is compli-cated and the subject of *petrology*. The most *refractory* magmas are those rich in Mg. A refractory substance is one that freezes at a high temperature and must be raised to a high temperature to melt. When Fe is added to the magma the melting point is lowered slightly because FeO is less refractory. High temperature magmas contain Fe, Mg, and other refractory elements like Titanium (Ti). High Mg basalts are the most refractory. When such magmas are flowing they are very fluid. As the Na, and K abundance increases the Fe may decrease. On Earth, magma bearing significant Na and/or K cools to form alkali basalt. But there are also volcanic rocks on Earth of the high alkali type that extruded from relatively high in the magma chamber at relatively low temperatures after the deeper magma chambers have cooled and crystalized removing much of the Fe and Mg from the magma mix. Such magmas may be rich in alkali metals such as Na, K, Lithium (Li), and low in Fe and Mg. The silica content of such magmas may be less than in typical Fe and Mg-bearing magmas, especially if the Al content is enriched. In this case alumino silicate lavas may result. As the aluminum and alkali abundances increase, the SiO_2 content will decrease (some of the silica combines with the aluminum) forming feldspathoid lavas that are more viscous, but may still flow for distances of many km across the surface. The more *viscous* feldspathoid lavas remain liquid at lower temperatures than the high temperature Mg-rich basalts. Table 10.1 lists the extrusion melting point, viscosity, and liquid density of some common lavas.

Other more or less exotic conditions may produce lava flows of compositions that have less common names and are of less common occurrence on Earth. Ash flows are also very fluid and, as mentioned in Section 10.6.1, may cover large surface areas.

Water in magma has two effects:

(1) If a hot magma is coming from a deep resevoir and contains considerable water it becomes very explosive when pressure is released suddenly, producing pyro-clastic cinders, beads and glasses.
(2) Water also reduces *viscosity* and is associated with volcanic lavas that flow long distances. A fluid medium to high temperature iron bearing magma is produced that can spread over considerable distances but still build a shield type volcano. Probably many lava flows on the flanks of the Olympus Mons volcano on Mars

Table 10.1. Melting points, viscosities, and denities of some common lavas.

Rock Type	Extrusive Melting Point (°C)	Extrusive Viscosity (Poise)	Liquid Density (g/cm3)
Lunar Basalt (~40% SiO_2)	1300–1400	10	2.7
Olivine Basalt (~45% SiO_2)	1100–1200	100–1000	2.7
Tholeiitic Basalt (~50% SiO_2)	1100–1200	1000–10,000	2.6
Andesites (~62% SiO_2)	1000–1100	10^6–10^7	2.4
Rhyolites (~75% SiO_2)	800–900	10^{10}–10^{11}	2.2

were formed by relatively water-rich basalts, perhaps of medium temperature coming from mid-level resevoirs, rich in iron. However, on the Moon the mare lavas are highly depleted in volatile elements and lack water. The very low viscosity of lunar basaltic lava (10 poise compared to 100–1000 poise for terrestrial olivine basalt) is the result of a relatively low abundance of SiO_2 (40%) and a high abundance of Fe, Ti and Mg, not water content.

10.6.3   What type of volcanism occurred on Mercury?

There is no morphological evidence for silica-rich volcanism on Mercury, nor is there any evidence for plate tectonics. The morphology of apparent volcanic plains is similar to flood basalts on the Moon and other terrestrial planets. This suggests that they were emplaced in a fluid condition. No large volcanic constructs are evident, as occur on Mars, Venus, and Earth, which further suggests the plains were erupted in large volumes from widely distributed sources, presumably fissures. From this evidence it is inferred that the plains are the Mercurian equivalent of flood basalts. Because of spectral evidence presented in Chapter 8, it is likely that basalts are of the Mg-rich or alkali-rich variety rather than the Fe-rich basalts common on Earth and the Moon. Spectroscopic measurements of Mercury's surface and of the thin atmosphere suggests that some of the smooth plains may be more alkali-rich than one would expect of a typical Mg-rich flood basalt especially around and in Caloris basis where enhancements of emissions from neutral potassium in the exosphere have been observed.

If the origin of both the intercrater and smooth plains is volcanic, then Mercury has experienced a period of volcanism much more extensive than that on the Moon. This would be consistent with thermal models that predict a thermal history with heating and melting a larger volume on Mercury than the Moon.

10.6.4   Compositions of Mercury's plains

As discussed in Chapter 8, telescopic observations of Mercury have found Mercury's surface is low or lacking in FeO. However, there is evidence for pyroxene (both orthopyroxene and clinopyroxene) and alkali elements. At first, these two observations may seem incompatible. But there are at least two ways in which these different types of volcanics could both be located and observed in the intercrater plains on Mercury. (1) There could have been an early volcanic event from deep within Mercury at the time the Fe was differentiating into the core of the planet with the Mg-rich fluid material covering great extents over Mercury's surface. Then in the late part of the heavy bombardment another volcanic episode could have begun where more shallow lavas, rich in alkalis and lower in SiO_2 and containing Ca-bearing pyroxene and feldspathoids were released in localized regions. Such basalts are called tephrite. (2) Another possibility is that following the early volcanism of Mg-rich basalts volcanism stopped. Near the end of the heavy bombardment or even later, excavation by impact cratering uncovered the highly differentiated alkali-rich material and exposed it at the surface in one or more locations. In fact, an anorthisitic crust may be overlain by basalts of varying composition. As mentioned earlier, *Mariner 10* recalibrated color ratioed images suggest Kuiper and Muraski craters may have excavated into such material. This material could be anorthosite. Anorthosite is a bulk rock formed from fractional crystallization as discussed earlier. It is composed of at least 90% plagioclase feldspar and up to 10% pyroxene. As we saw in Chapter 8, a spectrum from Mercury shows great similarity to a spectrum from a lunar breccia composed of 90% plagioclase feldspar and 10% pyroxene. The Mercury spectrum was from the intercrater plains. It appears from mid-IR spectroscopy that the feldspar in Mercury's intercrater plains may be more Na-rich than that of the lunar anorthosite which is generally Ca-rich. It is also interesting to note that when imaging the Na in Mercury's thin atmosphere, enhanced emission of Na was observed over the Kuiper/Muraski crater complex. The Na could be baking out of Na-bearing plagioclase feldspars that have been relatively recently excavated.

10.6.5   Mercury's smooth plains composition

Several mid-IR spectra have been obtained from 205–240° giving evidence for low Si and, perhaps mafic mineral content. This region is west of the smooth plains in and north of Tolstoj on the unimaged side of the planet. Other spectroscopic evidence for low Si content is just east of 15° longitude, again on the side unimaged by *Mariner 10*.

A low-Si (undersaturated in silica) and alkali-rich signature is consistant with feldspathoidal-bearing basalts and the high K abundance observed over the smooth plains in a near Caloris basin. Smooth plains are also formed from impact melt on the floors of craters formed in energetic impacts. The differentiation of these melts may have resulted in unusul compositions, especially if the target material was low in FeO and rich in alkalis. The earth-based spectra may, at least, in part represent these compositions.

The sophisticated spectrographs on the *MESSENGER* spacecraft will determine the chemical composition at high spatial resolutions. This, combined with other geological data, should greatly help characterize these plains units and finally answer the puzzling questions that remain regarding the geochemical make-up and history of Mercury's surface.

11

Tectonics

Crustal deformation (tectonics) occurs on Mercury, the other terrestrial planets, and the Moon. Crustal deformation is the result of stresses in the outer layers of a planet that are produced by thermal and/or mechanical processes. This deformation forms characteristic structures (faults and/or folds) that reveal the nature and direction of the stresses responsible for their formation. Mercury appears to have a tectonic framework which is unique in the Solar System.

11.1 FAULTING

In order to understand the causes and history of crustal deformation, three major characteristics must be determined: 1) the type of deformation must be identified. This can be accomplished by studying the morphology of the structures and the way they displace or deform the surface. 2) the distribution and orientation of the structures, and the time they formed must be determined. This procedure involves plotting the structures on a map, measuring their orientation, and observing their age relative to other landforms (for example, faults that cross craters or other terrain are younger than the craters and terrain. Conversely, if a fault is disrupted by a crater or is partially buried by plains, then it is older than those landforms). 3) the nature and distribution of the stresses that caused the deformation must be inferred from the types of structures, their distribution, and the pattern they form.

Unfortunately, not all of these characteristics can now be determined for Mercury. Since only about half the planet has been explored we do not know the global distribution of the tectonic features. To make matters worse, only about 25% of the surface was viewed at sun angles low enough to undertake terrain analysis. Therefore, tectonic features that occur in areas viewed at high sun angles are probably not discernable. So, in essence, only about 25% to 30% of the planet has been tectonically mapped. However, even this amount of mapping is sufficient to constrain certain tectonic models.

11.1.1   Fault types and mechanics

There are three basic types of faults: 1) normal, 2) thrust, and 3) transverse (Figure 11.1). Each is formed by a different stress regime. A convenient way to visualize these regimes is to consider the stresses as directed along three principal axes at right angles to each other. One axis is the maximum principal stress, another is the intermediate principal stress, and the third is the minimum principal stress. The orientation of these principal stress axes with respect to the surface determines the type of fault that forms. Faults form at an angle between the maximum and minimum stress axes and parallel the to intermediate stress axis. If the type of fault is known, then the orientation of the principal stresses can be inferred.

If the crust of a planet is stretched or pulled apart it is in tension. When this occurs the lateral or confining pressure is eased, and the weight of the rocks acting vertically exerts the greatest stress on the crust. In this case, the maximum stress axis is perpendicular to the surface, and the minimum and intermediate stress axes are parallel to the surface. If the maximum stress exceeds the strength of the rocks, then the crust will fracture. It will break along a fault plane that is inclined at a steep angle between the maximum and minimum stress axes and parallel to the intermediate stress axis. One block will move downward along a fault plane sloping toward the down-dropped block. This type of fault is called a normal fault. It is also called a tension fault because it forms when the crust is under tension. Another term for this type of fault is gravity fault because the maximum stress is vertical and due to the force of gravity acting on the rocks. Often this type of fault forms troughs where a section of the crust slides downward between two oppositely facing normal faults. In this case it is called a graben.

The opposite situation occurs when the crust is pushed together or compressed. Here the stress field is reversed with the maximum stress axis parallel to the surface and the minimum stress axis vertical. Again the crust breaks along a fault plane inclined at an angle between the maximum and minimum stress axes and parallel to the intermediate stress axis. In this case, however, one block is pushed or thrust over another block along a gently sloping fault plane that dips beneath the over-thrust block.

The third type of fault forms when both the maximum and minimum stress axes are parallel the surface and the intermediate stress axis is perpendicular. Again the fault plane forms at an angle between the maximum and minimum stress axes, but in this case the fault plane is vertical and one block slides past the other with little vertical movement. This type of fault is rare/absent on the Moon, Mars, and Mercury, rare on Venus, but relatively common on Earth where they form one of the plate boundaries called transform faults. The San Andreas fault of California is a well-known example of a transverse fault, that forms the boundary between the Pacific and North American tectonic plates.

In summary, normal faults result from tension that pulls the crust apart and

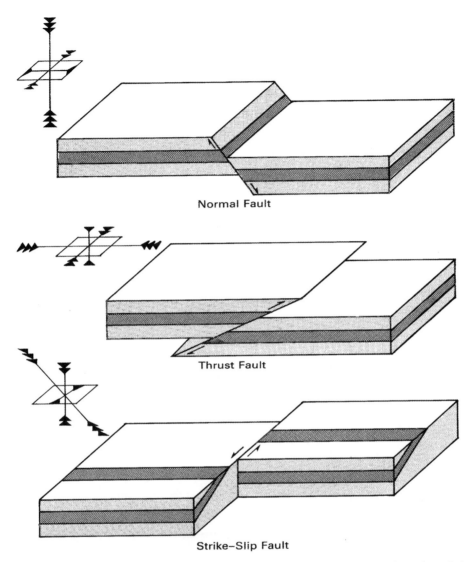

Figure 11.1. These diagrams show the three basic types of faults and the orientation of the stress field that causes them. Normal faults caused by tension are rare on Mercury, and strike–slip faults have not been confirmed. Thrust faults which are very common on Mercury form the lobate scarps. Normal faults represent crustal lengthening, thrust faults represent crustal shortening, and area is conserved for strike slip faults (from Strom, 1987).

causes crustal lengthening. Thrust faults are due to compression that pushes the crust together, causing crustal shortening. In transverse faulting two segments of the crust slide by each other and the surface area is conserved (neither crustal lengthening or shortening) (Figure 11.1).

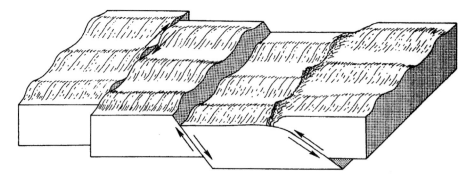

Figure 11.2. The three types of faults exhibit different surface expressions; see text for explanation (from Strom, 1987).

11.1.2   Topographic expression

Each type of fault has a different topographic expression. Normal faults or grabens form linear or arcuate scarps or troughs that vertically displace the topography they transect; their scarps are generally steep with sharp crests. Transverse faults are very straight, show little vertical relief, and laterally displace the landforms they transect. Thrust faults vertically displace the surface, but their crests are rounded because the fault plane dips beneath the overthrust block, producing an overhang that is unstable and collapses under its own weight. The surface trace is usually sinuous, because the fault plane slopes at a low angle and forms a meandering path as it cuts across different elevations. Thrust faults also push one portion of the surface over another so that landforms are shortened (Figure 11.2).

11.2   MERCURY'S LOBATE SCARPS

One of *Mariner 10*'s most important discoveries was the observation of long, sinuous cliffs or scarps that traverse Mercury's surface for hundreds of km. They were termed lobate scarps because they are characterized by rounded and lobed fronts that wind across the surface. Their lengths vary from about 20 km to over 500 km, and they can reach heights of about 1.5 km or more. Earth-based radar has determined heights of 700 m and widths of over 70 km for some of these scarps. Also, shadow, photoclinometry and several stereoscopic measurements have provided some good topographic determinations. The scarps transect a variety of terrain including craters, intercrater plains and smooth plains. Landforms that are transected by these scarps are displaced, indicating they are indeed faults. Similar faults occur on the Moon and other terrestrial planets.

11.2.1   Thrust faults

Mercury's lobate scarps have a surface expression exactly the same as that expected of thrust faults (Figures 11.3 and 11.4). They are sinuous scarps with rounded crests

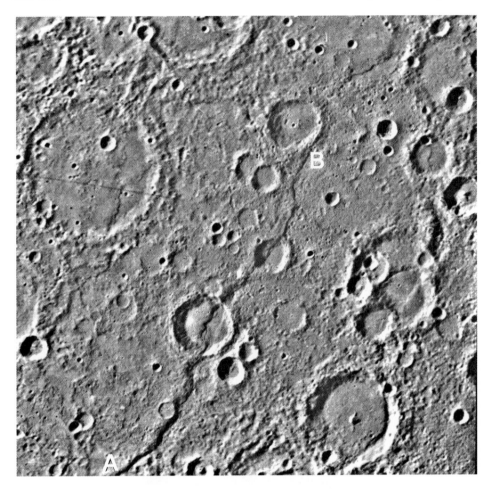

Figure 11.3. This is Discovery fault that traverses Mercury's surface for about 400 km from A to B. It is one of the largest viewed by *Mariner 10*, and reaches about 1.5 km high.

that in some cases greatly distort the craters they transect. One outstanding example is Guido d'Aresso crater whose rim is cut by a thrust fault named Vostok. The northeast part of this crater has been thrust over the southwestern part, causing a shortening of the crater's diameter and a 10 km horizontal offset of its rim (Figure 11.5).

11.2.2   Distribution and age

Where topographic analysis is possible, lobate scarps are found on all types of terrains and in almost all regions viewed by *Mariner 10*. They appear to be evenly distributed in the equatorial, mid-latitude and polar regions of the explored part of Mercury. Lobate scarps also show no preferred orientation, although there is a slight

Figure 11.4. This high-resolution image of Discovery fault was taken by *Mariner 10* on its third encounter with Mercury.

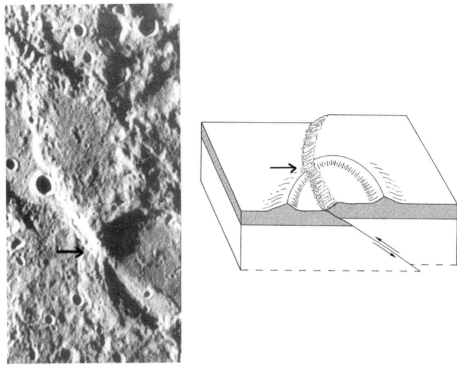

Figure 11.5. This image shows Vostok scarp (130 km long) displacing the rim of Guido d'Arezzo crater by about 10 km (arrow). The diagram on the right illustrates how the thrust fault has shortened the diameter leading to the horizontal offset (arrow). Guido d'Arezzo crater is 65 km diameter.

tendency for N 45° W and N 45° E directions. The more-or-less random distribution of thrust faults on the explored part of Mercury suggests that they may, in fact, be similarly distributed across the entire surface of the planet. If this turns out to be the case, then the planet experienced a period of global compression, and a net decrease in surface area since the compression began.

The time of formation of lobate scarps can be estimated from the age of the thrust faults relative to various terrain units. The scarps cut across the intercrater plains and all relatively old degraded craters in the highlands. There is no instance on the explored side where intercrater plains embay a thrust fault. Craters that disrupt faults are all relatively young with fresh-appearing morphology. These structural relationships suggest that the onset of scarp formation (compression and planet contraction) occurred after intercrater plains formation and at a time when the period of heavy bombardment was declining. If the period of heavy bombardment was a cataclysmic event, then scarp formation occurred after this event which ended about 3.8 billion years ago. The age of the scarps relative to smooth plains formation is more difficult to assess. They are very common on the plains within and surrounding the Caloris basin, but these scarps are probably related to subsidence of the basin floor and surroundings. The areas of smooth plains in the uplands appear to be somewhat deficient in lobate scarps compared to the surrounding terrains. However, these areas are rather small and it could be that the seeming paucity is not real. Future explorations will be needed to solve this problem. It appears that the tectonic framework represented by the lobate scarps began relatively late in Mercurian geologic history; possibly about 3.8 billion years ago.

11.2.3   The shrunken planet

With certain assumptions, the amount of decrease in surface area can be estimated from the total length and height of the scarps, and the inclination of the fault planes. The amount of horizontal displacement or crustal shortening along a thrust fault is simply the length of the fault multiplied by the vertical displacement divided by the tangent of the fault plane inclination. The lengths of the faults are readily measured. The vertical displacement is probably close to the heights of the scarps, but these heights are more difficult to measure. Scarp heights have been measured from the shadows they cast or by some Earth-based radar profiles. They appear to range in height from a few hundred meters to maybe as much as 2 km for Discovery Rupes. Furthermore, the faults often have a rounded crest which is higher than the height that is casting the shadow. In these cases the heights are a minimum. In the calculations it is assumed that the average height of the scarps is about 0.5–1 km. The inclination of the fault plane is considerably more difficult to estimate. On Earth thrust faults have fault plane inclinations ranging from about 45° (high-angle reverse faults) to about 25° for most thrust faults. However, for thrust faults the angles can vary widely and the faults can be imbricated. This means that they can be made up of a series of low-angle fault planes. In fact, the Vostok scarp appears to show several layers which could be individual fault planes of an imbricated thrust fault. The sinuous morphology of the Mercurian faults is much more compatible with thrust faults than high-angle reverse faults. Therefore, it is much more likely that the fault planes dip at angles closer to 25° than 45°. If it is assumed that the average fault plane inclination is 25° and that the average vertical displacement is from 500 m to 1 km, then the amount of crustal shortening can be estimated. It also must be assumed that the number and lengths of the faults measured on about 25% of the surface is representative of the other 75%. With these uncertainties in mind, the surface area lost because of crustal shortening is estimated to be about 31,000 to 63,000 km^2. This implies that Mercury's diameter has decreased by about 1–2 km since the onset of compression. Although there is much uncertainty involved in these estimates they are probably good to within about a factor of 2. The age of the faults suggests the global contraction began about 3.8 billion years ago. This places constraints on Mercury's thermal history and other tectonic models.

   If the assumptions used for these calculations, particularly the global distribution of thrust faults, prove to be true from future explorations, then Mercury is truly unique in the Solar System. No other planet or satellite has had its entire crust shortened and its diameter decreased from tectonic activity. The cause of this contraction is probably cooling of its crust and mantle, and, particularly, its enormous iron core.

11.2.4   Thrusting, lithospheric and crustal thickness

Mechanical modeling of Discovery Rupes gives some idea of the faulting parameters, and the thickness of the lithosphere at the time of faulting. Based on *Mariner 10* stereo images that cover the area of Discovery Rupes, the maximum height is about 1.5 km. Furthermore, there appears to be a parallel trough about 40 km wide located 90 km behind the scarp. The trough is interpreted to be a syncline that defines the dimension of the upper plate of the thrust fault. Mechanical modeling suggests that the fault plane inclination is about 30–35° with a displacement of about 2 km. The depth of faulting is estimated at about 35–40 km which should be the thickness of the lithosphere at the time of faulting ~3.8 billion years ago. If the limiting isotherm for Mercury's crust is about 300–600°C and it occurred at a depth of 40 km, then the corresponding heat flux at the time of faulting was about 10–40 W/m^2, which is less than old terrestrial oceanic lithosphere. Since Mercury has long-wavelength topography only certain combinations of crustal thickness and thermal structure are possible. This together with the faulting data suggests that the concentration of radiogenic elements in the crust and mantle of Mercury is at least 80% that Earth's value, and that the present thickness of the crust is about 100–200 km.

11.3   OTHER TECTONIC STRUCTURES

Although thrust faults dominate the tectonic framework of Mercury, there are other tectonic structures that are much less prominent or have a much more limited distribution.

11.3.1   Grabens and normal faults

At least the side viewed by *Mariner 10* is remarkably free of structures indicative of tension. Only the grabens and valleys (probably also grabens) on the floor of the Caloris basin and the jumbled terrain of the hilly and lineated terrain are due to tensional stresses. However both of these structural provinces are either the direct or indirect result of the Caloris basin impact. The grabens on the floor of the Caloris basin are probably due to an uplift of the floor following the formation of the interior plains and its ridges. The hilly and lineated terrain was caused directly by seismic waves generated by the Caloris basin impact and focused at the antipodal region (see Chapter 9).

11.3.2   Linear structures

Another less prominent set of linear features may have a tectonic origin. They consist of linear portions of crater rims and linear ridge-like structures. These lineations trend in two main directions, northeast and northwest, with a weaker north–south direction. This may represent an ancient fracture pattern formed very early in Mercury's history. The much younger valleys in the hilly and lineated terrain also trend in the same two main directions. They may have formed along fractures of this system when the region was disrupted by the Caloris impact-generated seismic waves.

The linear portions of crater rims may have formed by slumping or preferential excavation along zones of weakness associated with a pre-existing fracture pattern. These linear rims occur on more degraded older craters as well as the younger ones. This indicates that the fractures were present before the old craters, and, therefore, established very early in Mercurian history.

11.3.3   Despinning fault pattern

Mercury may have rotated much faster just after its formation and was subsequently slowed by tidal forces until it was captured into the present 3:2 spin–orbit resonance. During this time of rapid rotation centrifugal forces would have produced an asymmetric shape with flattened poles and bulging equator. As Mercury slowed and the centrifugal forces decreased, its shape would have become more spherical.

When changing from a flattened sphere (oblate spheroid) to a sphere, the bulging equatorial regions would have contracted and the flattened polar regions would have expanded proportionally (Figure 11.6). This process would have produced a stress pattern leading to a unique tectonic framework. The contracting equatorial region would have produced east–west directed compressional stress, with the expanding polar regions inducing north–south directed tensional stresses. In the equatorial regions north–south oriented thrust faults would form and east–west

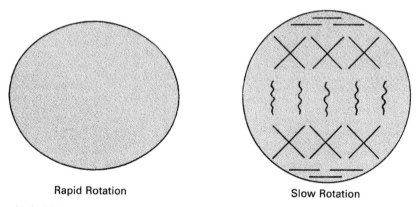

Rapid Rotation   Slow Rotation

Figure 11.6. This diagram shows the tectonic consequence of despinning Mercury from a higher rate. At a higher spin rate Mercury might have flattened poles as shown at the left. As the rotation slowed, it would have become more spherical, causing stresses above the fracture limit. The resulting tectonic framework is show in the right diagram where north–south thrust faults occur in the equatorial region, orthogonal strike–slip faults in mid-latitudes, and east–west normal faults in the polar regions. This type of tectonic framework is not seen on Mercury (from Strom, 1987).

directed normal faults would form in the polar regions. At mid-latitudes this interplay of stresses would produce northeast and northwest trending transverse fractures.

This mechanism, however, cannot account for the present tectonic framework of lobate scarps for four reasons. Firstly, the lobate scarps are just as abundant in the polar regions as in the equatorial regions. Secondly, no normal faults (tension) are found in the polar regions. Thirdly, the lobate scarps have a more-or-less random orientation, and, Fourthly, they are relatively young features that post-date the lineament system. However, the northeast and northwest systems of ancient lineations may be the remnants of a transverse fracture system caused by planetary despinning. The ancient age of this fracture pattern indicates that, if this despinning occurred, it did so very early in Mercury's history, probably before the formation of intercrater plains and many of the older craters.

12

History and origin

12.1 GEOLOGIC HISTORY

The geologic history of Mercury is uncertain. The earliest recorded events are the formation of intercrater plains about 4 billion years ago during the period of intense bombardment. These plains may have erupted through fractures caused by the intense bombardment on a relatively thin lithosphere, the upper part of which may have been composed of anorthosite. Near the end of the intense bombardment the Caloris basin was formed by a large impact that caused the hilly and lineated terrain from focused seismic waves in the antipodal region. Further eruption of lava within and surrounding the Caloris basin and other large craters and basins formed the smooth plains about 3.8 billion years ago. The system of thrust faults formed by global contraction after emplacement of the intercrater plains, but how soon after is unknown. If the thrust faults resulted from core cooling only, then they may have begun forming after smooth plains emplacement and resulted in about 1–2 km decrease in radius. As the core continued to cool and the lithosphere thickened, compressive stresses may have closed off the magma sources and volcanism may have ceased. Mercury's volcanic events probably took place early in its history, maybe in the first 700–800 million years. Since that time only occasional impacts of asteroids and comets took place. If the outer core is still in a molten condition, then Mercury may still be contracting as the outer core cools.

12.2 THERMAL HISTORY

Planetary thermal history models depend on compositional assumptions, such as the abundance of uranium (U), potassium (K), and thorium (Th), in the planet, as well as other physical parameters. Since our knowledge of Mercury's composition and physical characteristics are limited, these models can only provide a general idea of the thermal history for certain starting assumptions. Nevertheless, they provide useful insights into possible modes and consequences of thermal evolution.

It is generally assumed that all of the terrestrial planets started in a molten or semimolten condition because of high thermal input by the accretional process itself. Furthermore, based on the apparent measurement of an active diopolar magnetic field at Mercury, it is assumed that the outer core is still in a molten condition today. However, the magnetic field could be a remanant field, in which case there is no need to invoke a present-day outer molten core (see Chapter 5). Indeed, thermal models strongly indicate that Mercury's core should be completely solidified at the present time unless there was some way of maintaining high core temperatures throughout Mercury's history.

12.2.1 Scenario #1, an active dipole

First let us consider the scenario where the magnetic field is a presently active dipole, there is a molten outer core of unknown thickness, and it is molten because of the presence of sulfur (S). Starting with initially global molten conditions, a thermal history model by Gerald Shubert, with between 0.2 to 5% S in the core indicates that the total amount of planetary radius decrease due to cooling is from about 6 to 10 km depending on the amount of S. About 6 km of this contraction is solely due to mantle cooling during about the first billion years before the start of inner core formation. The amount of radius decrease due to inner core formation alone is negligible for 5% S and about 4 km for 0.2% S (Figure 12.1). This presents a real problem for the onset time and amount of global contraction inferred from the thrust faults.

If the maximum radius decrease of 2 km inferred from the thrust faults was due *solely* to cooling and solidification of the inner core, then the core S content is about 2 to 3% and the thickness of the outer fluid core is about 500 to 600 km. In this case, the inner core began to form about 3 billion years ago, after the period of intense bombardment. This implies that the tectonic framework began at about the same time, and that the smooth and intercrater plains were emplaced before this event. This is consistent with the geological evidence that indicates that at least the observed tectonic framework began to form after intercrater plains formation and possibly after smooth plains formation, but still in the first billion years after formation.

However, under initially molten conditions, the thermal history models indicate that the lithosphere has always been in compression, and that as much as a 6 km radius decrease occurred before inner core formation due to cooling of the mantle alone. This conflicts with the geologic history deduced from the origin and age of the plains and tectonic framework. If the smooth and intercrater plains are volcanic flows, then there must have been conduits for lavas to reach the surface and form the observed extensive deposits (intercrater plains are the most extensive terrain type on the observed portion of the planet). Early lithospheric compressive stresses could have slowed or stopped that process. However, the lithosphere may have been relatively thin (<50 km) at this time. Large impacts would probably result in extensive

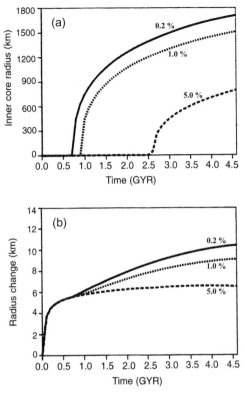

Figure 12.1. In (a) a thermal history model is shown for the inner core radius as a function of time for three values of initial core S content (0.2%, 1%, and 5%). Diagram (b) shows a model of the decrease in Mercury's radius due to mantle cooling and inner core growth for the same values of initial core S (from Schubert *et al.*, 1988).

fractures which may have provided easy egress for lavas to reach the surface and bury compressive structures. However, there is no geologic evidence for very early compressive stresses. At least some thrust faults that have been partially buried by intercrater plains should be present. None are observed on the explored hemisphere, but some could occur on the unexplored side.

    In contrast, thermal history models with initially cool conditions, iron (Fe) uniformly distributed throughout the planet, and heating by the decay of chondritic abundance of radioactive elements, indicate that core formation began about 1.2 billion years after accretion and was complete by 1.8 billion years after accretion. Such models lead to extensive melting and expansion that put the lithosphere in tension and lead to extensional fracturing providing easy egress for lavas (intercrater plains) to reach the surface. However, these events occur too late in Mercury's history to account for the inferred surface ages, and the model does not account for the large Fe core. Furthermore, it is doubtful that Mercury formed in an initially cool condition.

12.2.2   Scenario #2, a remanent field

If the magnetic field is a remanent field it loosens the constraint for a presently molten outer iron core. In this case, the core would be completely solid at present and its S content would be less than 0.2%. For a S content of 0.2%, the core takes almost the age of the Solar System (4.6 billion years) to solidify. With less it would solidify much faster, but it would still take well over a billion years to completely solidify because of its very large size. This small amount of S has implications for the origin of Mercury. The problem of the amount of radius decrease does not go away under this scenario. In fact it makes it worse. The amount of radius decrease would be at least 10 km which is not even close to the maximum estimate (2 km) based on the thrust faults. It is unlikely that there are significantly more thrust faults on the unexplored side than on the surface analyzed. Furthermore the onset of compressive stresses also occurs very early in Mercury's history. This alternative scenario produces a somewhat different history, but still presents difficulties with the geology.

12.2.3   No thermal model consistent with the geologic history

In short, there are distinct conflicts with thermal history models and the observed geology. These conflicts are surely due to our lack of data, and, therefore, lack of understanding of both the geological and geophysical processes on Mercury. Future spacecraft exploration of Mercury will hopefully supply the data to resolve these vexing conflicts.

12.3   ORIGIN OF PRESENT DAY MERCURY

Mercury's iron core is much larger in proportion to its rocky mantle than any other planet or satellite in the solar system. The primary problem concerning the origin of Mercury is how it acquired such a large iron core. We are fairly confident that the planets accreted from a solar nebula of gas and dust during the final stages of the Sun's formation. The large compositional difference between the outer jovian planets and the inner terrestrial planets was probably caused by a temperature gradient in this nebula. The inner part of the nebula was at high temperatures where *refractory material* predominated out to about $\sim 4\,\mathrm{AU}$, while the outer part beyond about $\sim 4\,\mathrm{AU}$ was at low temperatures where volatile condensates such as ices would predominate. Thus, the outer planets are large bodies consisting predominately of H and He with their satellites mostly consisting of a mixture of rocks with water and other ices. The inner planets are dominated by silicate mantles and crusts, and iron cores. The problem with Mercury is that its iron core is much larger than predicted by solar nebula *chemical equilibrium* condensation models.

12.3.1   Chemical equilibrium models

12.3.1.1   *Early condensation models*

Early chemical equilibrium condensation models that assumed Mercury formed entirely at its present distance from the Sun could not account for Mercury's large iron core. Furthermore, the models predicted almost the complete absence of S (100 parts per trillion of FeS). Other volatile elements and compounds, such as water, are also severely depleted (<1 part per billion of H). The absence of S is a severe problem if Mercury still contains an outer molten iron core.

12.3.1.2   *Turbulence broadened the possibilities*

More realistic models, where a significant part of Mercury is formed from material at more distant feeding zones, relax the S problem but still result in an uncompressed density of $4.3 \, g/cm^3$ rather that the observed $5.3 \, g/cm^3$. In other words, no chemical equilibrium condensation model seems to explain Mercury's enormous iron core.

12.3.2   How did the core get so large?

Three hypotheses have been proposed to account for Mercury's large iron core and to explain the discrepancy between the predicted and observed Fe abundance. One hypothesis (selective accretion) involves an enrichment of Fe due to mechanical and dynamical accretion processes in the innermost part of the solar nebula. The other two (post-accretional vaporization and giant impact stripping) invoke removal of a large fraction of the silicate mantle from a once larger proto-Mercury.

12.3.2.1   *Differential sorting of iron and silicates*

In the selective accretion model, the differential response of Fe and silicates to impact fragmentation and aerodynamic sorting leads to Fe enrichment owing to the higher gas density and the shorter dynamical time scales in the innermost part of the solar nebula. The removal process for silicates from Mercury's present position is more effective than that for Fe, leading to Fe enrichment. Selective accretion requires that Mercury accrete at about its present distance from the Sun in order for these dynamical processes to operate. In this model of the inner solar nebula must have lower temperatures than the equilibrium condensation model mentioned above.

12.3.2.2   *Post-accretion vaporization*

The post-accretion vaporization hypothesis proposes that intense bombardment by solar electromagnetic and corpuscular radiation in the earliest phases of the Sun's evolution (T-tauri phase) vaporized and drove off much of the silicate fraction of Mercury, leaving the core intact. The lost mantle and retention of the original core results in the large iron/silicate ratio and Mercury's high density.

12.3.2.3   *Giant impact stripping*

In the giant impact hypothesis, a planet-sized object impacts Mercury and essentially blasts away much of the planet's rocky mantle, leaving a core about the size of the original. This model is essentially the same as the impact model for the formation of the Earth, but in this case the impact is more direct and does not result in a satellite.

12.3.3   Which hypothesis is correct?

Discriminating between these models may be possible from knowledge of the chemical composition of the rocky silicate fraction. John Lewis estimates that for the selective accretion model, Mercury's silicate portion should contain about 3.6 to 4.5% alumina (Al), about 1% alkali oxides (Sodium (Na) and potassium (K)), and between 0.5 and 6% FeO. Post-accretion vaporization should lead to very severe depletion of alkali oxides ($\sim$0%) and FeO (0.1%) and extreme enrichment of refractory oxides ($\sim$40%). If a giant impact stripped away the crust and upper mantle late in accretion, then alkali oxides may be depleted (0.01 to 0.1%), with refractory oxides between about 0.1 to 1% and FeO between 0.5 and 6%. Unfortunately, our current knowledge of Mercury's silicate composition is extremely poor, but near- and mid-infrared (IR) spectroscopic measurements favor low FeO ($\sim$3%) and alkali-bearing feldspars or basalts with similar SiO_2 content. If the exosphere of Na and K is being outgassed from the crust, as seems possible from recent observations, than the post-accretional vaporization model may be unlikely. If FeO abundances are about 3 percent they are too high for this model. Deciding between the other models is not possible with our current state of knowledge about the crustal composition. Since selective accretion requires Mercury to have formed near its present position, then S should be nearly absent unless the solar nebula temperatures in this region were considerably lower than predicted by the chemical equilibrium condensation model. If there were lower temperatures then more FeO would also be supplied. If Mercury's magnetic field is a remanant field and the core is completely solidified, then the required very low core S abundance would be consistent with that model (Figure 12.2).

12.4   CLASHING PLANETS

Support for the giant impact hypothesis comes from three-dimensional computer simulations of terrestrial planet formation for several starting conditions. Since these simulations are by nature stochastic, a range of outcomes are possible (Figures 12.3, and 12.4).

They all suggest that significant fractions of the terrestrial planets may have accreted from material formed in widely separated parts of the inner Solar System. The simulations also suggest that Mercury may have experienced large excursions in its semi-major axis during its accretion. Proto-Mercury may have ranged over distances of 0.4 to 1.4 AU owing to energetic impacts during accretion (Figure 12.5). Consequently, Mercury could have accumulated material originally formed

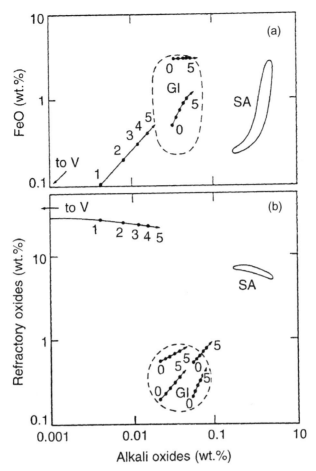

Figure 12.2. This diagram shows the possible bulk composition of the silicate mantle for the three models of Mercury's origin based on an analysis by John Lewis: selective accretion (SA), post-accretion vaporization (V), and giant impact (GI). The composition is parameterized for the FeO content, the alkali content (soda plus potash), and the refractory oxide content (Ca plus Al plus Ti oxides). The modifying effects of late infall of 0 to 5% meteorite material on several regolith compositions are indicated by arrows labeled 0 to 5 (from Lewis, 1988).

over the entire terrestrial planet range of heliocentric distances. About half of Mercury's mass could have accumulated from material originally formed at distances between about 0.8 and 1.2 AU (Figure 12.6). If so, then Mercury may have acquired a significant amount of S and other semi-volatiles like Na and K from material formed in regions of the solar nebula where S is stable. Plausible models estimate from 0.1 to 3% FeS abundance. However, the most extreme models of accretional mixing result in homogenizing the entire terrestrial planet region, contrary to the observed large systematic density differences.

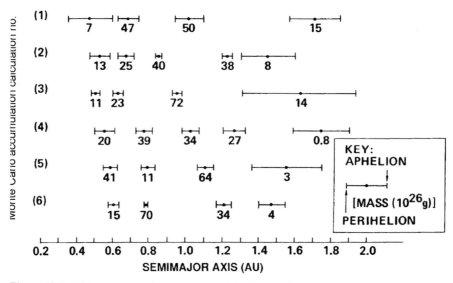

Figure 12.3. This shows the final outcome of six Monte Carlo accumulation calculations using five hundred 2×10^{25} g bodies. The semimajor axes of the final planets are indicated by points; the line through each point extends from the perihelion to the aphelion distances of the planet. The number under each point indicates the final masses of the bodies in units of 10^{25} g (from Wetherill, 1988).

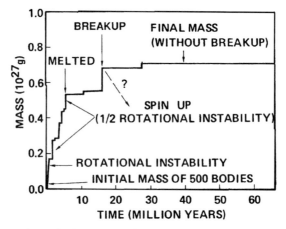

Figure 12.4. The growth on the innermost planet Mercury for case (1) in Figure 12.3 shows that the growth is punctuated by a number of giant impacts, one of which is energetic enough to cause major disruption of the planet (from Wetherill, 1988).

The computer simulations also indicate that the by-products of terrestrial planet formation are planet-sized objects up to three times the mass of Mars that become perturbed into eccentric orbits (mean $e \sim 0.15$ or larger). They eventually collide with the terrestrial planets during their final stages of growth. The final growth and giant

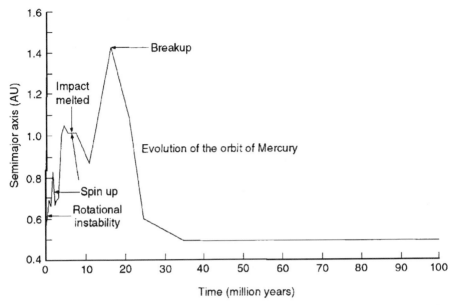

Figure 12.5. This shows the evolution of the semimajor axis of Mercury for case (1) in Figure 12.3. During the course of growth, the heliocentric distance of the body spans the entire terrestrial planet region (from Wetherill, 1988).

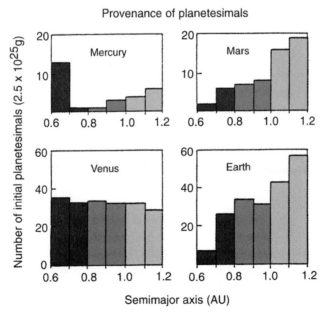

Figure 12.6. This diagram shows the source of the planetesimals as a function of semimajor axes that formed the final bodies for case (1) of Figure 12.3. A significant number of the planetesimals comes from feeding zones in the vicinity of Earth (from Wetherill, 1988).

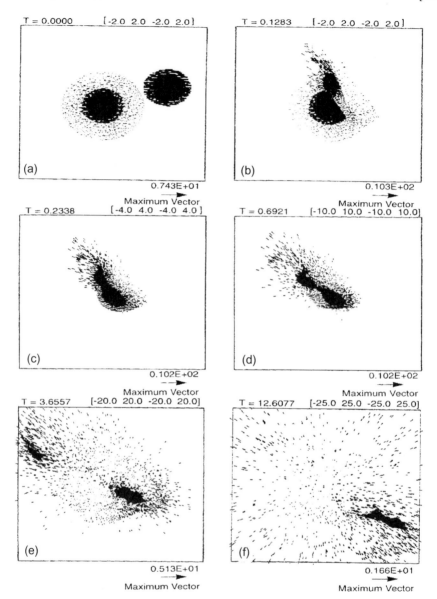

Figure 12.7. Computer simulation of a large off-axis impact with Mercury at 35 km/s by W. Benz. It shows that the mantle separates from the core. At least a portion of the mantle must re-accrete to form the present-day Mercury (from Benz, Slattery and Cameron, 1988).

impacts occur within the first 50 million years of Solar System history. Such large impacts may have resulted in certain unusual characteristics of the terrestrial planets, such as the slow retrograde rotation of Venus, the origin of the Moon, the Martian crustal dichotomy, and Mercury's large iron core.

   In at least one computer simulation where proto-Mercury grew to 2.25 times its present mass with an uncompressed density of about $4\,g/cm^3$, nearly central collisions of large projectiles with iron cores impacting at $20\,km/s$, or non central collisions at $35\,km/s$, resulted in a large silicate loss and little Fe loss (Figure 12.7). In the former case, although a large portion of Mercury's iron core is lost, an equally large part of the impactor's iron core is retained, resulting in approximately the original core size. At Mercury's present distance from the Sun, the ejected material re-accretes back into Mercury if the fragment sizes of the ejected material are greater than a few centimeters. However, if the ejected material is in the vapor phase or fine-grained ($<1\,cm$), then it will be drawn into the Sun by the *Poynting–Robertson effect* in a time shorter than the expected collision time with Mercury (about a million years). The proportion of fine-grained and large-grained material ejected from such a large impact is very uncertain. Therefore, it is not known if a large impact at Mercury's present distance would exclude enough mantle material to account for its large iron core. However, the disruption event need not have occurred at Mercury's present distance from the Sun (it could have occurred at a much greater distance, for example, $>0.8\,AU$ [see Figure 12.5]). In this case, the ejected mantle material would be mostly swept up by the larger terrestrial planets, particularly Earth and Venus.

13

Future exploration of Mercury

13.1 WE MUST RETURN TO MERCURY

It must be obvious by now that further exploration of Mercury is imperative in order to understand how the planet evolved. Fortunately, two missions are scheduled that should provide significant new data about the planet and solve many of the problems raised in this book; one is an American mission and the other is a joint European and Japanese mission. The American mission is called *MESSENGER* while the European/Japanese mission is called *Bepi Colombo* for a noted Italian celestial mechanician who discovered that *Mariner 10* could make multiple encounters with Mercury.

13.2 THE NASA *MESSENGER* MISSION

The name *MESSENGER* is an acronym for MErcury: Surface, Space ENvironment, GEochemisty, Ranging. To the ancients, Mercury was the messenger of the gods. This mission is one of NASA's Discovery series of planetary exploration missions that is managed by the Applied Physics Laboratory of Johns Hopkins University in Maryland, USA. The *MESSENGER* mission is a single spacecraft that will orbit Mercury and send back enormously important data from a variety of instruments over a period of at least one year. It is scheduled for launch in March 2004 and will orbit Mercury in April 2009 (Table 13.1). The reason it takes 5 years and 4 flybys before it is put in orbit is because it is necessary to use Venus and Mercury to slow the spacecraft enough to require only a relatively small retrorocket for the orbit insertion maneuver, and to be able to carry a relatively large payload (Figure 13.1, see colour section). *MESSENGER* will be placed in an elliptical orbit with a 200 km periapse altitude located at about 60° north latitude with a 12

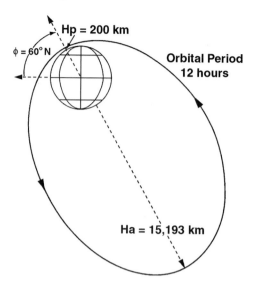

Figure 13.2. The orbital configuration of the *MESSENGER* spacecraft is shown in this diagram. It is an elliptical orbit with closest approach (Hp) of 200 km at 60° north latitude, and a greatest distance (Ha) of 15,193 km. The orbit will have a 12 hr period (Courtesy of *MESSENGER* Project, Applied Physics Laboratory, Laurel, MD).

Table 13.1. *MESSENGER* timeline of major events.

Event	Date
Launch	10 March 2004
1st Venus Encounter	24 June 2004
2nd Venus Encounter	16 March 2006
1st Mercury Encounter	21 July 2007
2nd Mercury Encounter	11 April 2008
Mercury Orbit Insertion	5 April 2009
End of Nominal Mission	4 April 2010

hour period (Figure 13.2). The spacecraft will also collect valuable data on its two preceding flybys of Mercury.

13.2.1   Mission objectives

There are seven main objectives of the mission, all of which are important to understanding the origin and evolution of Mercury and the inner planets. One is to determine the nature of the polar deposits including their composition. Another objective is to determine the properties of Mercury's core including its precise diameter, and whether or not it has a liquid component and, if so, its thickness. This will be accomplished by accurately measuring Mercury's libration amplitude from the laser altimeter and radio science experiments. Another objective is to determine variations in the structure of the lithosphere and whether or not convection is currently taking place. Determination of the nature of the magnetic field and whether it is currently on active dipole field, a remanent field, or both is still another objective. There are several instruments to study the chemical and mineralogical composition of the crust which should place constraints on Mercury's origin, and decide between the three competing hypotheses. Also these data will be extremely important to decipher Mercury's geology. The geologic evolution of Mercury will be addressed by the dual camera system which will image the entire surface at high-resolution and at a variety of wavelengths. Finally, the exosphere will be studied to determine its composition and interaction with the magnetosphere and surface (Figure 13.3, see colour section).

13.2.2 Science experiments

There are nine science experiments on board the spacecraft. They are as follows:

(1) a dual imaging system;
(2) a gamma-ray and neutron spectrometer;
(3) a magnetometer;
(4) a laser altimeter;
(5) atmospheric (0.155 to 0.6 µm) and surface (0.3 to 1.45 µm) spectrometers;
(6) an energetic particle and plasma spectrometer;
(7) an X-ray spectrometer;
(8) a laser altimeter; and
(9) a radio science experiment that uses the telecommunication system.

These instruments will be used to accomplish the objectives discussed in the preceding section. They are listed in Table 13.2 together with the measurements they will make. The *MESSENGER* website (http://messenger.jhuapl.edu) gives a comprehensive explanation of the *MESSENGER* mission, and Mercury information.

Table 13.2. *MESSENGER* instruments and measurements.

Instrument	Observation
Dual imaging system (1.5° and 10.5° FoV)	Surface mapping in stereo (12 filters)
Gamma ray and neutron spectrometer	Surface composition (O,Si,Fe,H,K)
X-ray spectrometer (1 to 10 KeV)	Surface composition (Mg,Al,Fe,Si,S,Ca,Ti)
Visible-infrared spectrograph	Surface composition
Ultraviolet–visible spectrometer	Exosphere composition
Magnetometer	Magnetic field
Laser altimeter	Topography of northern hemisphere
Energetic particle and plasma spectrometer	Energetic particles and plasma
Radio science (X-band transponder)	Gravity field and physical libration

13.3   THE EUROPE/JAPAN BEPI COLOMBO MISSION

The European Space Agency (ESA) has approved a "Cornerstone" mission to Mercury. It will provide more information on Mercury and its environment and will be an important complement to *MESSENGER*.

At present (2003) the Bepi Colombo mission is still being defined. However, it may consist of three spacecraft; two orbiters and one lander. The proposed launch dates are either 2011 or 2012. Possibly one orbiting spacecraft will be in a circular orbit at a periapse altitude of about 400 km (*MESSENGER*'s is elliptical with a periapse at 200 km), while the other is in a highly elliptical orbit to study the magnetic field and charged particle environment. A possible lander would make measurements of the surface at a high latitude for thermal considerations. The kinds of experiments and instruments have not yet been decided. The data returned from this mission will be an important complement to *MESSENGER*'s data and could fill in some gaps such as altimetric data for the southern hemisphere which *MESSENGER* will not measure.

Appendix A

Orbital and physical data for Mercury

ORBITAL DATA

Semimajor axis	0.3871 AU (5.79×10^7 km)
Perihelion distance	0.3075 AU (4.60×10^7 km)
Aphelion distance	0.4667 AU (6.98×10^7 km)
Sidereal period	87.97 days
Synodic period	115.88 days
Eccentricity	0.20563
Inclination	7.004°
Mean orbital velocity	47.87 km/s
Rotation period	58.646 days

PHYSICAL DATA

Radius	2,439 km
Surface area	7.475×10^7 km^2
Volume	6.077×10^{10} km^3
Mass	3.302×10^{26} g
Mean density	5.44 g/cm^3
Surface gravity	370 cm/s^2
Escape velocity	4.25 km/s
Surface temperatures	90 to 740 K ($-183°$ to 467°C)
Normal albedo (5° phase angle)	0.125
Magnetic dipole moment*	$4.9(\pm0.2) \times 10^{22}$ gauss cm^3

*If the magnetic field is a currently active dipole and not a remanent field

Appendix B

Glossary of terms

Absolute age. The age of a geologic unit measured in years.

Acceleration. Change in velocity with time. Either getting faster or getting slower or changing direction.

Accretion. The gradual accumulation of mass, as of a planet forming by the buildup of colliding particles in the solar nebula.

Achondrite. The smaller of two classes of stony meteorites containing no chondrules (see chondrites).

Albedo. The ratio of the amount of light reflected from a planet or other body to the amount of light it receives from the Sun.

Altimetric profiles. Plots of the altitudes above a reference height of the surface features of a planet or satellite.

Antipodal point. The point exactly on the opposite side of a planet (i.e., 180° apart on a sphere).

Apparition. The appearance of a planet in the sky where it is easily viewed from Earth.

Aphelion. Point in an orbit at which a planet or other body is farthest from the Sun.

Apollo program. The American program to land humans on the Moon (1961–1972).

Asteroid. One of several tens of thousands of small planets ranging in size from a few hundred kilometers to less than 1 km in diameter, with an orbit generally between the orbits of Mars and Jupiter but also in other orbits throughout the solar system.

Asthenosphere. The partially melted or weak layer underlying the outer rigid lithospheric layer of a planet or satellite.

Astronomical unit (AU). The average distance between the Earth and the Sun (about 150 million kilometers).

Aurora. Light radiated by atoms and ions in the ionosphere, mostly in the magnetic polar regions.

Ballistic range. The distance an ejected particle travels before striking the ground.

Ballistic trajectory. The looping path that an ejected particle travels while in flight.

Basalt. A fine-grained dark volcanic rock with mineral components including feldspar, olivine, and pyroxene.

Blue-shift. The term used to describe the phenomenon of an apparent shift to shorter wavelength of either a sound or light owing to the observer's relative motion toward the source (or the relative motion of the source toward the observer, or both).

Bow shock wave. Boundary where the solar wind encounters the magnetic field of a planet.

Breccia. Jagged rock and mineral fragments cemented together.

Caldera. A large volcanic depression formed by collapse caused by the withdrawal or eruption of underlying magma.

Cassegrain telescope. A reflecting telescope in which the secondary mirror reflects the light back through a hole in the primary mirror to a point behind the primary mirror.

Central peak. A peak created in the center of an impact crater at the time of formation when the shock wave generated during the impact reflects back from the solid rock beneath causing an uplift.

Chemical equilibrium. For a chemical reaction the set of conditions at which the forward and reverse reaction rates are equal.

Chondrite. The most numerous type of stoney meteorite named for its content of ovoid spherical millimeter sized masses of silicates (chondrules).

Collisionally evolved. A size distribution, different from the original one, resulting from mutual collisions over a long period of time.

Collisionally unevolved. A size distribution not appreciably affected by collisions since the original formation of the population.

Commensurability. The state in which two periods (e.g., rotation and orbital, or orbital periods of two objects) are whole number multiples of each other.

Complex crater. Craters with central peaks, interior wall terraces and flat floors.

Continuous ejecta blanket. The hummocky surface immediately surrounding an impact crater, consisting of excavated material.

Continuum. The average radiance in a spectrum that represents the level from which the absorptions or emissions can be measured.

Copernican theory. The heliocentric, or sun-centered view of the Solar System.

Coronal mass ejection. Solar material from huge erupting bubbles moving away from the sun ahead of erupting prominences.

Cosmic rays. Charged particles (atomic nuclei and electrons) moving in space at close to the speed of light.

Cross-tail current. Electrons and ions streaming in opposite directions across the magnetotail creating electrical currents.

Crust. The relatively thin, outermost layer of a planet. This layer is chemically distinct from the underlying mantle.

Declination. The north–south angular measurement of a celestial object on the grid of the celestial sphere, like latitude on the angular grid covering Earth and with $0°$ defined to be the great circle above the Earth's equator.

Depolarized. Radar signals having lost their original orientation after reflection from a surface.

Despinning. The slowing down of a body's rotation rate due to the action of tidal forces.

Differentiation. A separation or segregation of different kinds of material in different layers in the interior of a planet or in a rock melt.

Dipole. Any object that is oppositely charged at two points.

Discontinuous ejecta. The region beyond the continuous ejecta blanket surrounding a crater, where strings and clusters of secondary craters and rays occur.

Distribution coefficient. A number that characterises the ability of one substance to distribute in another.

Doppler shift. The apparent lengthening or shortening of a wave because of motion either away or toward the source of the wave (sound or light).

Eastern elongation. The eastern-most angular distance of a planet from the Sun as measured on the celestial sphere; for an inferior planet resulting in an evening apparition.

Eccentricity. A parameter that specifies the shape of an elliptical orbit. The ratio of the distance between the foci of an ellipse and the major (or longest) axis; circles have an eccentricity of zero.

Ecliptic plane. The plane of the Earth's orbit about the Sun.

Ejecta. Material excavated during the formation of an impact crater and deposited around the crater.

Ejecta blanket. The debris ejected from an impact composed of material from both the surface and the impactor that forms a continuous cover over the surrounding terrain.

Elongation. A planet's angular distance from the Sun.

End member component. One of the pure substances of two or more substances commonly forming a solid solution.

Enstatite. A type of ortho-pyroxene characterized by magnesium (Mg).

Ephemeris. A tabulation of the positions of a celestial object in an orderly sequence for a number of dates.

Escape velocity. The speed an object must achieve in order to break away gravitationally from another body.

Excavation cavity. A crater formed by the ejection of material.

Flare (solar flare). A rapid brightening of a region around a sunspot.

Fraction illuminated. The fraction of a planet's or satellite's illumined disk that is visible to an observer.

Fractional crystallization. The stepwise cooling and separation of crystals from the parent magma melt; crystals formed at higher temperatures are generally denser than the melt and fall to the bottom of the melt chamber while crystals forming at cooler temperatures are less dense and remain at the top.

Fraunhofer lines. Absorptions in the Sun's visible spectrum caused by cool gases overlying hotter gases of the photosphere.

Galilean satellites. The four largest satellites of Jupiter.

Geomagnetic. Referring to the Earth's magnetic field.

Graben. A fault trough in which a section of the crust slides downward between two oppositely facing normal faults.

Granite. A coarse-grained igneous rock containing quartz and potassium–aluminum silicates.

Gravity assist. The technique of using a planet's gravitational attraction to change the speed and direction of a spacecraft without using fuel.

Gyro. A wheel or disk mounted to spin rapidly about an axis that is free to turn in various directions but which resists any motion that would change the axial direction of spin.

Harmonic Law. Kepler's third law of planetary motion: the cubes of semimajor axes of the planetary orbits are proportional to the squares of the sidereal periods of the planets' revolutions about the Sun.

Heavy bombardment. The period of time, apparently between 3.8 to 4.5 billion years ago, when the cratering rate was high throughout the Solar System.

High-gain antenna. The antenna on a spacecraft which has a parabolic reflector for the purpose of concentrating the energy of electromagnetic radiation either transmitted to, or received from the Earth.

Hot poles. The subsolar points on Mercury at 0 and 180° W longitude that face the Sun at perihelion, so named because they are the hottest locations on the planet.

Hydrazine. A corrosive, fuming liquid NH_2NH_2 (or N_2H_4) that is used for fuel on rockets and spacecraft.

Hydrodynamic dynamo. The mechanism thought to generate magnetic fields at the Earth, and perhaps Mercury, characterized by a conducting outer core, liquid enough to undergo convection and move differentially over and around the solid inner core thus maintaining the magnetic field.

Inclination (of an orbit). The angle between the orbital plane of a body and the ecliptic plane.

Inferior conjunction. The configuration in which an inferior planet lies exactly between the Earth and the Sun.

Inferior planet. A planet whose orbit lies between that of the Earth and the Sun.

Infrared radiometry. The study of the thermal properties of a distant object by gathering and measuring the infrared radiation emanating from that object.

Intercrater plains. Level to gently rolling surfaces with a rough texture due to a large number of small superposed craters. They occur between clusters of craters in heavily cratered areas.

Interplanetary magnetic field (IMF). The magnetic field throughout the Solar System resulting from the solar wind.

Interplanetary medium. The sparse distribution of gas, electric, and magnetic fields, and solid particles in the space between the planets.

Ionosphere. The upper region of an atmosphere in which many of the atoms are ionized by ultraviolet sunlight and interaction with ionized particles.

Isotope. Any of two or more forms of the same element whose atoms all possess the same number of protons but different numbers of neutrons.

Kepler's Laws. Three laws, discovered by Johannes Kepler, that describe the motions of the planets.

Launch window. A range of dates during which a space vehicle can be launched for a specific mission without exceeding the fuel capabilities of that system.

Law of Equal Areas. Kepler's second law: the radius vector from the Sun to any planet sweeps out equal areas in the planets' orbital plane in equal intervals of time.

Limb. The edge of a planetary body as viewed from a different, distant body.

Lithosphere. The outer rigid layer of a planet or satellite.

Lobate scarps. Gently sinuous cliffs resulting from thrust faults formed by compressive stresses.

Low-gain antenna. The antenna on a spacecraft that transmits and receives from all directions at once.

Magma. Subsurface molten rock from which igneous rocks are formed.

Magnetic axis. The imaginary line along which a magnetic dipole points.

Magnetopause. The inside shock zone of the magnetosheath encountered around a magnetized planet or satellite.

Magnetosheath. An elongated region defined by interaction of charged particles in the solar wind and the planet's magnetic field; it forms a cavity in which the planet is protected from direct impact of the solar wind.

Magnetosphere. The region around a planet occupied by the planet and its magnetic field.

Magnetotail. The name for the region of the magnetosphere streaming behind the magnetized body interacting with the solar wind to make the magnetosphere.

Mantle. The middle layer of a planet, between the crust and the core, and chemically distinct from them.

Mare (maria, pl.). Latin for "sea", name applied to most large dark regions on the Moon.

Maturity. A term used to describe the cumulative effect of space weathering on a planetary surface; a very old, strongly weathered surface is mature.

Megaregolith. Discovered at the Moon, a deep layer (kilometers deep) of broken rocks formed from aeons of meteorite bombardment (see regolith).

Micrometeorites. Meteorites only a few microns (10^{-4} cm) in diameter.

Morphology. The shape and structure of a landform.

Node. The intersection of the orbit of a body with the ecliptic plane.

Normal fault. A fault resulting from tension that pulls the crust apart and causes crustal lengthening.

Obliquity. The tilt of a planet's rotation axis to its orbital plane.

Occultation. An eclipse of a star, planet, or spacecraft by a satellite or a planet.

Orbital plane. The plane defined by a planet's or satellite's orbit about the parent body and the point at the center of the parent body.

Perihelion. The closest approach of a planet to the Sun.

Petrogesis. Having to do with the origins of the composition of rocks.

Phase. The ratio of the illuminated portion of a planetary or satellite disk to the entire illuminated disk (the circumference is taken to be a circle).

Phase angle. The angle measured at the central point of a body between the source of illumination and the location of the observer; thus the body observed is at the vertex of the phase angle.

Phase change. The change in physical condition of a material from one form to another, for example liquid to solid.

Photometry. The measurement of light intensities.

Photon. A small, quantized amount of light energy.

Planetesimals. Bodies of intermediate size, most of which finally accreted to larger bodies.

Plasma sheet. A region of the magnetopause where charged particles flow; generally in a plane perpendicular to and midway between the magnetic poles.

Plate tectonics. The motion of plates of the lithosphere relative to one another.

Poynting–Robertson effect. An effect of radiation pressure from sunlight that causes small, micro-sized particles orbiting the Sun to spiral inward toward the Sun.

Precession. A slow rotation of a planet's axis due to the gravitation effects of a larger body on the planet's equatorial bulge.

Quadrature. A planetary elongation 90° east or west of the Sun.

R Plot also called *Relative Plot.* A graph that shows the plots of number of craters on a surface for a range of diameter sizes with data displayed in a specified format.

Radiation acceleration. The gain in velocity of a body caused by the impact of photons on the body.

Radiation pressure. The force per unit area on a body caused by the impact of photons on the body.

Radiation zones. Regions about a planet or satellite where magnetic fields keep ionized particles entrapped (see Van Allen radiation belt).

Ray system. Bright elongated streaks emanating outward from a fresh impact crater.

Red-shift. The term used to describe the phenomenon of an apparent shift to longer wavelength of either a sound or light wave owing to the observer's relative motion away from the source (or the relative motion of the source away from the observer, or both).

Reflectance. A measure of the ratio of the reflected light from a surface to the incident light on the surface.

Refractory material. A substance that is resistant to heat (i.e., does not melt easily).

Regolith. The upper most layer of an airless planetary, satellite, or asteroid surface composed of the broken rocks, rock powders, interplanetary dust accumulations, meteoritic impact melts, and breccias.

Remanent field. A magnetic field that remains after the period of implantation of the magnetism.

Resonance scattering. In airglow studies the process whereby a photon is absorbed by an atom and then later emitted at the same wavelength.

Ring effect. In light reflected from a planetary surface, a complicated light scattering process that has the result of lessening the depth of Fraunhofer lines in the solar spectrum.

Rotational speed. The angular speed of a point on the surface of a planet or satellite owing to the rotation of that body.

Rupes. Latin for scarps or cliffs.

Seismic waves. Vibrations traveling through a planet's interior that result from earthquakes or impacts.

Shield volcano. A broad, gently sloping volcano built by flows of fluid basaltic lava.

Sidereal period. The period for one object to complete an orbit around another body.

Silicates. Rocks and minerals that are composed mostly of oxygen and silicon in the form of silica (SiO_2).

Simple crater. A bowl shaped crater.

Size–frequency distribution. A graphic illustration depicting the variation in the number of craters at various diameters.

Solar nebula. A cloud of gas and dust from which the Solar System is presumed to have formed.

Solar wind. A stream of charged particles, mostly protons and electrons, that escape the Sun's outer atmosphere at high speeds and stream out into the Solar System.

Solid solutions. The solid equivalent of dissolving one or more substances in another such as salt in water.

Space weather. Fluctuations and rapid changes in the average state of the solar wind and interplanetary electric and magnetic fields.

Space weathering. The sum total of all the effects of the solar wind and solar radiation field on a planet's or satellite's surface.

Spectral slope. The rise or fall in the brightness value of a spectrum as the wavelength increases; usually in the context of visible and near-infrared spectra from surfaces of planets, satellites, or asteroids.

Spin–orbit coupling. A relationship between a planet's rotational and orbital periods such that they are whole number multiples of each other.

Stratigraphy. The characteristics of the layers of rock strata and their interpretation in terms of origin and history.

Subsidence zones. Regions of a planetary surface that have dropped relative to the surrounding regions.

Superior planet. A planet that is farther from the Sun than the Earth.

Synchronous period. A situation in which a planet or satellite always keeps the same face toward the body around which it revolves.

Synodic period. The interval between successive similar lineups of a body with the Sun.

Tectonics. The study of the large-scale movements and deformation of a planet's crust.

Telluric absorptions. A term used in remote sensing from Earth to denote absorptions in spectra coming from outside of Earth's atmosphere (i.e. from other planets or stars), caused by atoms and molecules in Earth's atmosphere.

Tensional rarefaction wave. The region of low density in a sound wave propagating through a solid that was caused by a sudden extension of the rock layer or layers.

Terminator. The line of sunrise or sunset on a body.

Thermophysical. A property of matter that characterizes the thermal properties of the substance.

Thrust fault. A fault resulting from compression that pushes part of the crust over another part and causes crustal shortening.

Transit. The passage of a planet (or satellite) across the face of the Sun (or parent planet).

Transparency. The property of surface materials that characterizes how well electromagnetic radiation penetrates the material.

TNT. The explosive compound Trinitrotoluene.

Ultraviolet light. Electromagnetic radiation of wavelengths shorter than the shortest (violet) wavelengths to which the eye is sensitive but longer than X-rays; usually defined to be between 100–4000 Å or 1.0×10^{-8}–4.00×10^{-7} m.

Ultraviolet spectroscopy. The measurement of ultraviolet light over an relatively large interval of wavelengths at many wavelength points giving a measurement in the relative radiance coming to the detector at each measurement point and resulting in a spectrum of the light measured.

Valles. Latin for valleys.

Van Allen radiation belt. A doughnut-shaped region surrounding the Earth where many rapidly moving charged particles are trapped in its magnetic field.

Vidicon camera. A camera that contains an electron gun and photoconductor permitting the beam of electrons emitted from the gun and collected by the photoconductor to be transformed into current that can subsequently be converted to a television image.

Viscosity. A liquid's resistance to flowing.

Volatile. A material easily vaporized.

Volume scattering. The term used to describe the interaction of electromagnetic radiation within individual crystals in a planetary regolith.

Voyagers. A series of spacecraft that were launched by the United States in 1977 to explore the outer Solar System.

Warm poles. The subsolar points on Mercury at 90 and 270° that face the Sun at aphelion. Although also on the equator they are cooler than the "hot poles".

Western elongation. The western-most angular distance of a planet from the Sun as measured on the celestial sphere; for inferior planets resulting in morning apparitions.

Wrinkle ridges. Sinuous ridges that occur on the lava plains of the Moon, Mars, Venus and Mercury, probably caused by compression.

Zenith. The point directly overhead.

Appendix C

Names and Locations of Mercury's Surface Features

CRATERS

Name	Latitude (°)	Longitude (°)	Diameter (km)
Abu Nuwas	17.4N	20.4W	116.0
Africanus Horton	51.5S	41.2W	135.0
Ahmad Baba	58.5N	126.8W	127.0
Al-Akhtal	59.2N	97.0W	102.0
Alencar	63.5S	103.5W	120.0
Al-Hamadhani	38.8N	89.7W	186.0
Al-Jahiz	1.2N	21.5W	91.0
Amru Al-Qays	12.3N	175.6W	50.0
Andal	47.7S	37.7W	108.0
Aristoxenus	82.0N	11.4W	69.0
Asvaghosa	10.4N	21.0W	90.0
Bach	68.5S	103.4W	214.0
Balagtas	22.6S	13.7W	98.0
Balzac	10.3N	144.1W	80.0
Barma	41.3S	162.8W	128.0
Bartók	29.6S	134.6W	112.0
Basho	32.7S	169.7W	80.0
Beethoven	20.8S	123.6W	643.0
Belinskij	76.0S	103.4W	70.0
Bello	18.9S	120.0W	129.0
Bernini	79.2S	136.5W	146.0
Bjornson	73.1N	109.2W	88.0

(*continued*)

Name	Latitude (°)	Longitude (°)	Diameter (km)
Boccaccio	80.7S	29.8W	142.0
Boethius	0.9S	73.3W	129.0
Botticelli	63.7N	109.6W	143.0
Brahms	58.5N	176.2W	96.0
Bramante	47.5S	61.8W	159.0
Brontë	38.7N	125.9W	60.0
Bruegel	49.8N	107.5W	75.0
Brunelleschi	9.1S	22.2W	134.0
Burns	54.4N	115.7W	45.0
Byron	8.5S	32.7W	105.0
Callicrates	66.3S	32.6W	70.0
Camões	70.6S	69.6W	70.0
Carducci	36.6S	89.9W	117.0
Cervantes	74.6S	122.0W	181.0
Cézanne	8.5S	123.4W	75.0
Chaikovskij	7.4N	50.4W	165.0
Chao Meng-Fu	87.3S	134.2W	167.0
Chekhov	36.2S	61.5W	199.0
Chiang K'ui	13.8N	102.7W	35.0
Chong Ch'ol	46.4N	116.2W	162.0
Chopin	65.1S	123.1W	129.0
Chu Ta	2.2N	105.1W	110.0
Coleridge	55.9S	66.7W	110.0
Copley	38.4S	85.2W	30.0
Couperin	29.8N	151.4W	80.0
Darío	26.5S	10.0W	151.0
Degas	37.4N	126.4W	60.0
Delacroix	44.7S	129.0W	146.0
Derzhavin	44.9N	35.3W	159.0
Despréz	80.8N	90.7W	50.0
Dickens	72.9S	153.3W	78.0
Donne	2.8N	13.8W	88.0
Dostoevskij	45.1S	176.4W	411.0
Dowland	53.5S	179.5W	100.0
Dürer	21.9N	119.0W	180.0
Dvorák	9.6S	11.9W	82.0
Echegaray	42.7N	19.2W	75.0
Eitoku	22.1S	156.9W	100.0
Equiano	40.2S	30.7W	99.0

Fet	4.9S	179.9W	24.0
Flaubert	13.7S	72.2W	95.0
Futabatei	16.2S	83.0W	66.0
Gainsborough	36.1S	183.3W	100.0
Gauguin	66.3N	96.3W	72.0
Ghiberti	48.4S	80.2W	123.0
Giotto	12.0N	55.8W	150.0
Gluck	37.3N	18.1W	105.0
Goethe	78.5N	44.5W	383.0
Gogol	28.1S	146.4W	87.0
Goya	7.2S	152.0W	135.0
Grieg	51.1N	14.0W	65.0
Guido d'Arezzo	38.7S	18.3W	66.0
Hals	54.8S	115.0W	100.0
Han Kan	71.6S	143.8W	50.0
Handel	3.4N	33.8W	166.0
Harunobu	15.0N	140.7W	110.0
Hauptmann	23.7S	179.9W	120.0
Hawthorne	51.3S	115.1W	107.0
Haydn	27.3S	71.6W	270.0
Heine	32.6N	124.1W	75.0
Hesiod	58.5S	35.0W	107.0
Hiroshige	13.4S	26.7W	138.0
Hitomaro	16.2S	15.8W	107.0
Holbein	35.6N	28.9W	113.0
Holberg	67.0S	61.1W	61.0
Homer	1.2S	36.2W	314.0
Horace	68.9S	52.0W	58.0
Hugo	38.9N	47.0W	198.0
Hun Kal	0.5S	20.0W	1.5
Ibsen	24.1S	35.6W	159.0
Ictinus	79.1S	165.2W	119.0
Imhotep	18.1S	37.3W	159.0
Ives	32.9S	111.4W	20.0
Janáček	56.0N	153.8W	47.0
Jókai	72.4N	135.3W	106.0
Judah Ha-Levi	10.9N	107.7W	80.0
Kalidasa	18.1S	179.2W	107.0
Keats	69.9S	154.5W	115.0

(continued)

Name	Latitude (°)	Longitude (°)	Diameter (km)
Kenko	21.5S	16.1W	99.0
Khansa	59.7S	51.9W	111.0
Kosho	60.1N	138.2W	65.0
Kuan Han-Ch'ing	29.4N	52.4W	151.0
Kuiper	11.3S	31.1W	62.0
Kurosawa	53.4S	21.8W	159.0
Leopardi	73.0S	180.1W	72.0
Lermontov	15.2N	48.1W	152.0
Lessing	28.7S	89.7W	100.0
Li Ch'ing-Chao	77.1S	73.1W	61.0
Li Po	16.9N	35.0W	120.0
Liang K'ai	40.3S	182.8W	140.0
Liszt	16.1S	168.1W	85.0
Lu Hsun	0.0N	23.4W	98.0
Lysippus	0.8N	132.5W	140.0
Ma Chih-Yuan	60.4S	78.0W	179.0
Machaut	1.9S	82.1W	106.0
Mahler	20.0S	18.7W	103.0
Mansart	73.2N	118.7W	95.0
Mansur	47.8N	162.6W	100.0
March	31.1N	175.5W	70.0
Mark Twain	11.2S	137.9W	149.0
Martí	75.6S	164.6W	68.0
Martial	69.1N	177.1W	51.0
Matisse	24.0S	89.8W	186.0
Melville	21.5N	10.1W	154.0
Mena	0.2S	124.4W	52.0
Mendes Pinto	61.3S	17.8W	214.0
Michelangelo	45.0S	109.1W	216.0
Mickiewicz	23.6N	103.1W	100.0
Milton	26.2S	174.8W	186.0
Mistral	4.5N	54.0W	110.0
Mofolo	37.7S	28.2W	114.0
Molière	15.6N	16.9W	132.0
Monet	44.4N	10.3W	303.0
Monteverdi	63.8N	77.3W	138.0
Mozart	8.0N	190.5W	270.0
Murasaki	12.6S	30.2W	130.0
Mussorgskij	32.8N	96.5W	125.0
Myron	70.9N	79.3W	31.0

Nampeyo	40.6S	50.1W	52.0
Nervo	43.0N	179.0W	63.0
Neumann	37.3S	34.5W	120.0
Nizami	71.5N	165.0W	76.0
Okyo	69.1S	75.8W	65.0
Ovid	69.5S	22.5W	44.0
Petrarch	30.6S	26.2W	170.0
Philoxenus	8.7S	111.5W	90.0
Pigalle	38.5S	9.5W	154.0
Po Chü-I	7.2S	165.1W	68.0
Po Ya	46.2S	20.2W	103.0
Polygnotus	0.3S	68.4W	133.0
Praxiteles	27.3N	59.2W	182.0
Proust	19.7N	46.7W	157.0
Puccini	65.3S	46.8W	70.0
Purcell	81.3N	146.8W	91.0
Pushkin	66.3S	22.4W	231.0
Rabelais	61.0S	62.4W	141.0
Rajnis	4.5N	95.8W	82.0
Rameau	54.9S	37.5W	51.0
Raphael	19.9S	75.9W	343.0
Ravel	12.0S	38.0W	75.0
Renoir	18.6S	51.5W	246.0
Repin	19.2S	63.0W	107.0
Riemenschneider	52.8S	99.6W	145.0
Rilke	45.2S	12.3W	86.0
Rimbaud	62.0S	148.0W	85.0
Rodin	21.1N	18.2W	229.0
Rubens	59.8N	74.1W	175.0
Rublev	15.1S	156.8W	132.0
Rudaki	4.0S	51.1W	120.0
Rude	32.8S	79.6W	75.0
Rumi	24.1S	104.7W	75.0
Sadi	78.6S	56.0W	68.0
Saikaku	72.9N	176.3W	88.0
Sarmiento	29.8S	187.7W	145.0
Sayat-Nova	28.4S	122.1W	158.0
Scarlatti	40.5N	100.0W	129.0
Schoenberg	16.0S	135.7W	29.0
Schubert	43.4S	54.3W	185.0

(continued)

Name	Latitude (°)	Longitude (°)	Diameter (km)
Scopas	81.1S	172.9W	105.0
Sei	64.3S	89.1W	113.0
Shakespeare	49.7N	150.9W	370.0
Shelley	47.8S	127.8W	164.0
Shevchenko	53.8S	46.5W	137.0
Sholem Aleichem	50.4N	87.7W	200.0
Sibelius	49.6S	144.7W	90.0
Simonides	29.1S	45.0W	95.0
Sinan	15.5N	29.8W	147.0
Smetana	48.5S	70.2W	190.0
Snorri	9.0S	82.9W	19.0
Sophocles	7.0S	145.7W	150.0
Sor Juana	49.0N	23.9W	93.0
Soseki	38.9N	37.7W	90.0
Sotatsu	49.1S	18.1W	165.0
Spitteler	68.6S	61.8W	68.0
Stravinsky	50.5N	73.5W	190.0
Strindberg	53.7N	135.3W	190.0
Sullivan	16.9S	86.3W	145.0
Sur Das	47.1S	93.3W	132.0
Surikov	37.1S	124.6W	120.0
Takanobu	30.8N	108.2W	80.0
Takayoshi	37.5S	163.1W	139.0
Tansen	3.9N	70.9W	34.0
Thakur	3.0S	63.5W	118.0
Theophanes	4.9S	142.4W	45.0
Thoreau	5.9N	132.3W	80.0
Tintoretto	48.1S	22.9W	92.0
Titian	3.6S	42.1W	121.0
Tolstoj	16.3S	163.5W	390.0
Ts'ai Wen-Chi	22.8N	22.2W	119.0
Ts'ao Chan	13.4S	142.0W	110.0
Tsurayuki	63.0S	21.3W	87.0
Tung Yüan	73.6N	55.0W	64.0
Turgenev	65.7N	135.0W	116.0
Tyagaraja	3.7N	148.4W	105.0
Unkei	31.9S	62.7W	123.0
Ustad Isa	32.1S	165.3W	136.0
Valmiki	23.5S	141.0W	221.0
Van Dijck	76.7N	163.8W	105.0

Van Eyck	43.2N	158.8W	282.0
Van Gogh	76.5S	134.9W	104.0
Velázquez	37.5N	53.7W	129.0
Verdi	64.7N	168.6W	163.0
Vincente	56.8S	142.4W	98.0
Vivaldi	13.7N	85.0W	213.0
Vlaminck	28.0N	12.7W	97.0
Vyasa	48.3N	81.1W	290.0
Wagner	67.4S	114.0W	140.0
Wang Meng	8.8N	103.8W	165.0
Wergeland	38.0S	56.5W	42.0
Whitman	41.4N	110.4W	70.0
Wren	24.3N	35.2W	221.0
Yeats	9.2N	34.6W	100.0
Yun Son-Do	72.5S	109.4W	68.0
Zeami	3.1S	147.2W	120.0
Zola	50.1N	177.3W	80.0

PLANITIA (PLAINS)

Name	Latitude (°)	Longitude (°)
Borealis Planitia	73.4N	79.5W
Budh Planitia	22.0N	150.9W
Caloris Planitia	30.5N	189.8W
Odin Planitia	23.3N	171.6W
Sobkou Planitia	39.9N	129.9W
Suisei Planitia	59.2N	150.8W
Tir Planitia	0.8N	176.1W

RUPES (LOBATE SCARPS)

Name	Latitude (°)	Longitude (°)
Adventure Rupes	65.1S	65.5W
Astrolabe Rupes	42.6S	70.7W
Discovery Rupes	56.3S	38.3W
Endeavour Rupes	37.5N	31.3W
Fram Rupes	56.9S	93.3W

(*continued*)

Name	Latitude (°)	Longitude (°)
Gjöa Rupes	66.7S	159.3W
Heemskerck Rupes	25.9N	125.3W
Hero Rupes	58.4S	171.4W
Mirni Rupes	37.3S	39.9W
Pourquoi-Pas Rupes	58.1S	156.0W
Resolution Rupes	63.8S	51.7W
Santa María Rupes	5.5N	19.7W
Victoria Rupes	50.9N	31.1W
Vostok Rupes	37.7S	19.5W
Zarya Rupes	42.8S	20.5W
Zeehaen Rupes	51.0N	157.0W

VALLIS (VALLEYS)

Name	Latitude (°)	Longitude (°)
Arecibo Vallis	27.5S	28.4W
Goldstone Vallis	15.8S	31.7W
Haystack Vallis	4.7N	46.2W
Simeiz Vallis	13.2S	64.3W

DORSA (RIDGES)

Name	Latitude (°)	Longitude (°)
Antoniadi Dorsum	25.1N	30.5W
Schiaparelli Dorsum	23.0N	164.1W

MONTES (MOUNTAINS)

Name	Latitude (°)	Longitude (°)
Caloris Montes	39.4N	187.2W

Bibliography

Baker, D.N. (1990) Energy coupling in the magnetospheres of Earth and Mercury. *Adv. Space. Res.*, **S**, 23–26.

Barlow, N.G., Allen, R.A., and Vilas, F. (1999) Mercurian impact craters, implications for polar ground ice. *Icarus*, **141**, 194–204.

Baumgardner, J., Mendillo, M., and Wilson, J.K. (2000) A digital high definition imaging system for spectral studies of extended planetary atmospheres, 1. Initial result in white light showing features on the hemisphere of Mercury unimaged by Mariner 10. *Astron. J.*, **119**, 2458–2464.

Benz, W., Slattery, W.L. and Cameron, A.G.W. (1988). Collisional Stripping of Mercury's Mantle. *Icarus*, **74**, 516–528.

Bida, T.A., Killen, R.M., and T.H. Morgan (2000). Discovery of Calcium in Mercury's Atmosphere, *Nature*, **404**, 159–161.

Binder, A.B., and Cruikshank, D.P. (1967) Mercury, new observations of the infrared bands of carbon dioxide. *Science*, **155**, 1135.

Blewett, D.T., Lucey, P.G., Hawke, B.R., Ling, G.G., and Robinson, M.S. (1997) A comparison of Mercurian reflectance and spectral quantities with those of the Moon. *Icarus*, **129**, 217–231.

Broadfoot, A.L., Kumar, S., Belton, M., and McElroy, M.B. (1974) Mercury's atmosphere from Mariner 10, preliminary results. *Science*, **185**, 166–169.

Broadfoot, A.L., Shemansky, D.E., and Kumar, S. (1976) Mariner 10: Mercury atmosphere. *Geophys. Res. Lett.*, **3**, 577–580.

Burns, J.O., Gisler, G.R., Borovsky, J.E., Baker, D.N., and Zeilik, M. (1987) Radio-interferometric imaging of the subsurface emissions from the planet Mercury. *Nature*, **329**(6136), 224–326.

Butler, B., Muhleman, D., Slade, M., and Jurgens, R. (1991) Mercury Goldstone/VLA Radar, Part II. *Bull. Amer. Ast. Soc.*, **23**(3), 1200.

Butler, B., Muhleman, D., and Slade, M. (1993) Mercury, full-disk radar images and the detection and stability of ice at the north pole. *J. Geophys. Res.*, **98**, 15003–15023.

Butler, B.J. (1997) The migration of volatiles on the surfaces of Mercury and the Moon. *J. Geophys. Res.*, **102**(E8), 19283–19291.

Cameron, A.G.W. (1985). The partial vaporizaton of Mercury. *Icarus*, **64**, 285–294.

Cameron, A.G.W., Fegley, B., Benz, W., and Slattery, W.L. (1988) The strange density of Mercury: theoretical considerations, In: F. Vilas, C. Chapman, and M. Matthews (eds), *Mercury*, pp. 692–705. University of Arizona Press, Tucson, Arizona.

Cheng, A.F., Johnson, R.E., Krimigis, S.M., and Lanzerotti, L.J. (1987) Magnetosphere, exosphere, and surface of Mercury. *Icarus*, **71**, 430–440.

Christon, S.P. (1987) A comparison of the Mercury and Earth magnetospheres, electron measurements and substorm time scales. *Icarus*, **71**, 448–471.

Christon, S.P., Feynman, J., and Slavin, J.A. (1987) Dynamic substorm injections, similar magnetospheric phenomena at Earth and Mercury. In: A. Lui (ed.), *Magnetotail Physics*. Johns Hopkins University Press, Baltimore, Maryland.

Cintala, M.J., Wood, C.A., and Head, J.W. (1977) The effects of target characteristics on fresh crater morphology. *Proc. 8th Lunar Sci. Conf.*, p. 3409.

Clark, P.E., Leake, M.A., and Jurgens, R.F. (1988) Goldstone radar observations of Mercury, In: F. Vilas, C. Chapman, and M. Matthews (eds), *Mercury*, pp. 77–100. University of Arizona Press, Tucson, Arizona.

Connerney, J.E.P., and Ness, N.F. (1988) Mercury's magnetic field and interior, In: F. Vilas, C. Chapman, and M. Matthews (eds), *Mercury*, pp. 429–460. University of Arizona Press, Tucson, Arizona.

Cooper, B.L, Potter, A.E, Killen, R., and Morgan, T.H. (2001) Mid-infrared spectra of Mercury. *J. Geophys. Res.*, **106**, 32803–32814.

Cooper, B.L., Salisbury, J.W., Killen, R., and Potter, A.E. (2002) Mid-Infrared spectral features of rocks and their powders. *J. Geophys. Res.*, **107**, 10.1029/2000JE001462.

Cuzzi, J.N. (1974) The nature of the subsurface of Mercury from microwave observations at several wavelengths. *Ap. J.*, **189**, 577–586.

Dantowitz, R.F., Teare, S.W., and Kozubal, M.J. (2000) Ground-based high-resolution imaging of Mercury. *Astron. J.*, **119**, 2455–2457.

Davies, M.E., Dwornik, S.E., Gault, D.E., and Strom, R.G. (1978) *Atlas of Mercury*, NASA, SP-423, National Aeronautics and Space Administration, US Government printing office, Washington, DC.

Domingue, D.L., Sprague, A.L., and Hunten, D.M. (1997) Dependence of mercurian atmospheric column abundance estimations on surface reflectance modeling. *Icarus*, **128**, 75–82.

Dunne, J.A., and Burgess, E. (1978) The Voyage of Mariner 10, NASA SP-424, Washington, DC.

Emery, J.P., Sprague, A.L., Witteborn, F.C., Colwell, J.E., Kozlowski, R.W.H., and Wooden, D.H. (1998) Mercury: thermal modeling and mid-infrared (5–12 micrometer) Observations, *Icarus*, **136**, 104–123.

Feldman, W.C., Barraclough, B.L., Hansen, C.J., and Sprague, A.L. (1997) The neutron signature of Mercury's volatile polar deposits. *J. Geophys. Res.*, **102**(E11), 25565–25574.

Gault, D.E., Guest, J.E., Murray, J.B., Dzurisin, D., and Malin, M.C. (1975) Some comparisons of impact craters on Mercury and the Moon. *J. Geophys. Res.*, **80**, 2444.

Gault, D.E., Burns, J.A., Cassen, P., and Strom, R.G. (1977) Mercury. *Ann. Reviews Astron. Astrophysics*, **15**, 97.

Gehrels, T., Landau, R., and Coyne, G.V. (1987) Mercury, wavelength and longitude dependence of polarization. *Icarus*, **71**, 386–396.

Gold, R.E., Solomon, S.C., McNutt, R.L., Santo, A.G., Abshire, J.B., Acuña, M.H., Afzal, R.S., Anderson, B.J., Andrews, F., Evans, L.G., Feldman, W.C., Follas, R.B. Gloecklen, G., Goldsten, J.O., Hawkins, S.E.III, Izenberg, N.R., Jasulek, S.E., Ketchum, E.A. Lankton, M.R., Lohr, D.A., Mauk, B.H., McClintock, W.E., Murchie, S.L., Schlemm,

C.E.II, Smith, D.E., Starr, R.D., Zurbuchen, T.H. (2001) The MESSENGER mission to Mercury: scientific payload. *Planetary and Space Science*, **49**, 1467–1479.

Goldstein, B.D., Suess, S.T., and Walker, R.J. (1981) Mercury, magnetospheric processes and the atmospheric supply and loss rate. *J. Geophys. Res.*, **86**, 5485–5499.

Harder, H., and Schubert, G. (2001) Sulfur in Mercurys core? *Icarus*, **151**, 118–122.

Harmon, J.K. (1997) Mercury radar studies and lunar comparisons. *Adv. Space Res.*, **19**, 1487–1496.

Harmon, J.K., and Slade, M.A. (1992) Radar mapping of Mercury: full-disk images and polar anomalies, *Science*, **258**, 640–642.

Harmon, J.K., Perillat, P.J., and Slade, M.A. (2001) High-resolution radar imaging of Mercury's north pole. *Icarus*, **149**, 1–15.

Hood, L.L., and Schubert, G. (1979) Inhibition of solar wind impingement on Mercury by planetary induction currents. *J. Geophys. Res.*, **84**, 2641–2647.

Hood, L.L., Zakharian, A., Halekas, J., Mitchell, D.L., Lin, R.P., Acuna, M.H., and Binder, A.B. (2001) Initial mapping and interpretation of lunar crustal magnetic anomalies using Lunar Prospector magnetometer data. *J. Geophys. Res.*, **106**, 27825–27839.

Hughes, H.G., App, F.M., and McGetchin, T.R. (1977) Global seismic effects of basin-forming impacts. *Phys. Earth Planet Interiors*, **15**, 251.

Hunten, D.M., and Sprague, A.L. (1997) Origin and character of the Lunar and Mercurian atmospheres. *Adv. Space Res.*, **19**(10), 1551–1560.

Hunten, D.M., and Sprague, A.L. (2002) Diurnal variation of Na and K at Mercury. *Meteoritics & Planetary Science*, **37**, 1191–1195.

Hunten, D.M., and Wallace, L.V. (1993) Resonance scattering by mercurian sodium. *Astrophys. J.*, **417**, 757–761.

Hunten, D.M., Morgan, T.H., and Shemansky, D. (1988) The Mercury atmosphere. In: F. Vilas, C.R. Chapman, and M.S. Matthews (eds), *Mercury*, pp. 562–612. Univ. of Arizona Press, Tucson, Arizona.

Ip, W.H. (1986) The Sodium exosphere and magnetosphere of Mercury. *Geophys. Res. Lett.*, **13**, 423–426.

Ip, W.H. (1987) Dynamics of electrons and heavy ions in Mercury's magnetosphere. *Icarus*, **71**, 441–447.

Ip, W.H. (1987) Mercury's magnetospheric irradiation effect on the surface. *Geophys. Res. Lett.*, **14**, 1191–1194.

Ip, W.H. (1990) On solar radiation-driven surface transport of sodium atoms at Mercury. *Astrophys. J.*, **356**, 675–681.

Ip, W.H. (1993) On the surface sputtering effects of magnetospheric charged particles at Mercury. *Ap. J.*, **418**, 451–456.

Jeanloz, R., Mitchell, D.L., Sprague, A.L., and Pater, I.D. (1995) Evidence for a basalt-free surface on Mercury and implications for internal heat. *Science*, **268**, 1455–1457.

Kabin, K., Gombosi, T.I., DeZeeuw, D.L., and Powell, K.G. (2000) Interaction of Mercury with the solar wind. *Icarus*, **143**, 397–406.

Killen, R.M. (1988) Resonance scattering by sodium in Mercury's atmosphere I. The effect of phase and atmospheric smearing. *Geophys. Res. Lett.*, **15**(1), 80–83.

Killen, R.M. (1989) Crustal diffusion of gases out of Mercury and the Moon. *J. Geophys. Lett.*, **16**(2), 171–174.

Killen, R.M., and Ip, W.H. (1999) The surface-bounded atmospheres of Mercury and the Moon. *Rev. Geophys.*, **37**, 361–406.

Killen, R.M., and Morgan, T.H. (1993) Maintaining the Na atmosphere of Mercury. *Icarus*, **101**, 293–312.

Killen, R.M., Morgan, T.H., and Potter, A.E. (1990) Spatial distribution of sodium vapor in the atmosphere of Mercury. *Icarus*, **85**, 145–167.

Killen, R.M., Benkhoff, J., and Morgan, T.H. (1997) Mercury's polar caps and the generation of an OH exosphere. *Icarus*, **125**(1), 195–211.

Killen, R.M, Potter, A.E., Fitzsimmons, A., and Morgan, T.H. (1999) Sodium D2 line profiles: clues to the temperature structure of Mercury's exosphere. *Plan. Space Sci.*, **47**, 1449–1458.

Killen, R.M., Potter, A.E., Reiff, P., Sarantos, M., Jackson, B.V., Hick, P., and Giles, B. (2001) Evidence for space weather at Mercury. *J. Geophys. Research*, **106**, 20509–20526.

Kumar, S. (1976) Mercury's atmosphere, a perspective after Mariner 10. *Icarus*, **28**, 579–591.

Leake, M.A. (1981) The intercrater plains of Mercury and the Moon: Their nature, origin and role in terrestial planet evolution. Ph.D. Dissertation, Dept. of Planetary Sciences, University of Arizona.

Ledlow, M.J. *et al.* (1992) Subsurface emissions from Mercury, VLA radio observations at 2 and 6 centimeters. *Ap. J.*, **384**, 640–655.

Lewis, J.S. (1988) Origin and composition of Mercury, In: F. Vilas, C. Chapman, and M. Matthews (eds), pp. 651–666, *Mercury*. Univ. of Arizona Press, Tucson, Arizona.

Madey, T.E., Yakshinskiy, B.V., Ageev, V.N., and Johnson, R.E. (1998) Desorption of alkali atoms and ions from oxide surfaces, relevance to origins of Na and K in atmospheres of Mercury and the Moon. *J. Geophys. Res.*, **103**(E3), 5873–5887.

Mallama, A., Wand, D., and Howard, R.A. (2002) Photometry of Mercury from SOHO/LASCO and Earth: The Phase function from 2 to 170 degrees. *Icarus*, **155**, 253–264.

McCord, T.B., and Adams, J.B. (1972) Mercury: interpretation of optical observations, *Icarus*, **17**, 585–588.

McCord, T.B., and Clark, R.N. (1979) The Mercury soil: presence of Fe2+. *J. Geophys. Res.*, **84**, 7664–7668.

McGrath, M.A., Johnson, R.E., and Lanzerotti, L.J. (1986) Sputtering of sodium on the planet Mercury. *Nature*, **323**, 696–696.

Melosh, H.J., and McKinnon, W.B. (1988) The tectonics of Mercury, In: F. Vilas, C. Chapman, and M. Matthews (eds), *Mercury*, pp. 374–400. Univ. of Arizona Press, Tucson, Arizona.

Mendillo, M.W.J., Limaye, S., Baumgardner, J., Sprague, A., and Wilson, J. (2001) Imaging the surface of Mercury using ground-based telescopes. *Planetary and Space Science*, **49**(14–15), 1501–1505.

Mitchell, D., and Pater, I.d. (1994) Microwave imaging of Mercury's thermal emission at wavelengths from 0.3 to 20.5 cm. *Icarus*, **110**, 2–32.

Morgan, T.H., and Killen, R.M. (1997) A non-stoichiometric model of the composition of the atmospheres of Mercury and the Moon. *Planet. Space Sci.*, **45**(1), 81–94.

Morgan, T.H., Zook, H.A., and Potter, A.E. (1988) Impact-driven supply of sodium and potassium in the atmosphere of Mercury. *Icarus*, **74**, 156–170.

Moroz, V.I. (1965) Infrared spectrum of Mercury (1.0–3.9 microns). *Sov. Astron. A.J.*, **8**, 882–889.

Moses, J.I., Rawlins, K., Zahnle, K., and Dones, L. (1999) External sources of water for Mercury's putative ice deposits. *Icarus*, **137**, 197–221.

Murray, B., Belton, M.J.S., Danielson, G.E., Davies, M.E., Gault, D.E., Hapke, B., O'Leary, B., Strom, R.G., Suomi, V., and Trask, N.J. (1974) Mariner 10 pictures of Mercury: first results. *Science*, **184**, 459.

Murray, B., Belton, M.J.S., Danielson, G.E., Davies, M.E., Gault, D.E., Hapke, B., O'Leary,

B., Strom, R.G., Suomi, V., and Trask, N.J. (1974) Mercury's surface: preliminary description and interpretation from Mariner 10 pictures, *Science*, **185**, 169.

Murray, B.C., Strom, R.G., Trask, N.J., and Gault, D.E. (1975) Surface history of Mercury: implications for the terrestrial planets, *J. Geophys. Res.*, **80**, 2508.

Murray, B. and Burgess, E. (1977) Flight to Mercury, Columbia University Press, New York, NY.

Nash, D.B. (1991) Infrared Reflectance Spectra (2.2 to 15 μm) of lunar samples. *Geophysical Res. Letters*, **18**, 2145–2147.

Ness, N.F., Behannon, K.W., Lepping, R.P., Whang, Y.C., and Schatten., K.H. (1974) Observations at Mercury encounter by the plasma science experiment on Mariner 10. *Science*, **185**, 159–170.

Nimmo, F. (2002) Constraining the crustal thickness on Mercury from viscous topographic relaxation. *Geophys, Res. Lt.*, **29**(5), 10.1029/2001GL013883.

Pike, R.J. (1988) Geomorphology of impact craters on Mercury, In: F. Vilas, C. Chapman, and M. Matthews (eds), *Mercury*, pp. 165–273. Univ. of Arizona Press, Tucson, Arizona.

Potter, A.E. (1995) Chemical sputtering could produce sodium vapor and ice on Mercury. *Geophys. Res. Lett.*, **22**, 3289–3292.

Potter, A.E., and Morgan, T.H. (1985) Discovery of sodium in the atmosphere of Mercury. *Science*, **229**, 651–653.

Potter, A.E., and Morgan, T.H. (1986) Potassium in the atmosphere of Mercury. *Icarus*, **67**, 336–340.

Potter, A.E., and Morgan, T.H. (1987) Variation of sodium on Mercury with solar radiation pressure. *Icarus*, **71**, 472–477.

Potter, A.E., and Morgan, T.H. (1990) Evidence for magnetospheric effects on the sodium atmosphere of Mercury. *Science*, **248**, 835–838.

Potter, A.E., and Morgan, T.H. (1997a) Sodium and potassium atmospheres of Mercury. *Planet. Space Sci.*, **45**(1), 95–100.

Potter, A.E., and Morgan, T.H. (1997b) Evidence for suprathermal sodium on Mercury. *Adv. Space Res.*, **19**(10), 1571–1576.

Rava, B., and Hapke, B. (1987) An analysis of the Mariner 10 color ratio map of Mercury. *Icarus*, **71**, 387–429.

Robinson, M.S., and Lucey, P.G. (1997) Recalibrated Mariner 10 color mosaics, implications for mercurian volcanism. *Science*, **275** (10 January), 197–200.

Robinson, M.S., and Taylor, G.J. (2001) Ferrous oxide in Mercury's crust and mantle. *Meteoritics & Planetary Science*, **36**, 841–847.

Russell, C.T., Baker, D.N., and Slavin, J.A. (1988) The magnetosphere of Mercury, In: F. Vilas, C. Chapman, and M. Matthews (eds), *Mercury*, pp. 514–561. Univ. of Arizona Press, Tucson, Arizona.

Salisbury, J.W., Walter, L.S., and D'Aria, D. (1988) Mid-infrared (2.5–13.5 μm) Spectra of Igneous Rocks Open-Files Report 88–686. USGS, Reston, Virginia.

Salisbury, J.W., Walter, L.S., Vergo, N., and D'Aria, D.M. (1991) Inrared (2.1–25 μm) spectra of minerals. Johns Hopkins University Press, Baltimore, MD. pp. 267.

Sandner, W. (transl) (1963) *The Planet Mercury*. Macmillan, New York, 93 pp.

Santo, A.G., Gold, R.E., McNutt, R.L., Solomon, S.C., Ercol, C.J., Farguhan, R.W., Hartka, T.J., Jenkins, J.E., McAdams, J.V., Mosher, L.E., Persons, D.F., Artis, D.A., Bokulic, R.S., Conde, R.F., Dakermunji, G., Goss, M.E., Haley, D.R., Heeres, K.J., Maurer, R.H., Moore, R.C., Rodberg, Elliot, H., Stern, T.G., Wileg, S.R., Williams, B.G., Yen, C.L., and Peterson, M.R. (2001) The MESSENGER mission to Mercury: spacecraft and mission design. *Planetary and Space Science*, **49**, 1481–1500.

Schubert, G., Ross, M.N., Stevenson, D.J., and Spohn, T. (1988) Mercury's thermal history and the generation of its magnetic field, In: F. Vilas, C. Chapman, and M. Matthews (eds), *Mercury*, pp. 429–460. Univ. of Arizona Press, Tucson, Arizona.

Schultz, P.H. (1988) Cratering on Mercury: a relook, In: F. Vilas, C. Chapman, and M. Matthews (eds), *Mercury*, pp. 274–335. Univ. of Arizona Press, Tucson, Arizona.

Sitko, M., Sprague, A.L., and Lynch, D.K. (eds) (2000) *Thermal Emission Spectroscopy and Analysis of Dusts, Disks and Regoliths*, pp. 187–196. Astronomical Society of the Pacific Conference Series, San Francisco.

Slade, M., Butler, B., and Muhleman, D. (1992) Mercury radar imaging, evidence for polar ice. *Science*, **258**, 635–640.

Smith, E.I., and Hartnell, J.A. (1979) Crater size-shape profiles for the Moon and Mercury: terrain effects and interplanetary comparisons. *Moon and Planets*, **19**, 479.

Smyth, W.H. (1986) Nature and variability of Mercury's sodium atmosphere. *Nature*, **323**, 696–699.

Smyth, W.H., and Marconi, M.L. (1995) Theoretical overview and modeling of the sodium and potassium atmospheres of Mercury. *Ap. J.*, **441**, 839–864.

Solomon, S.C. (1977) The relationship between crustal tectonic evolution in the Moon and Mercury. *Phys. Earth Planet. Interiors*, **15**, 135.

Solomon, S.C., McNutt, R.L., Gold, R.E., Acuña, M.H., Baker, D.N., Boynton, W.V., Chapman, C.R., Cheng, A.F., Gloeckler, G., Head, J.W.III., Krimigis, S.M., McClintock, W.E., Murchie, S.L., Peale, S.J., Phillips, R.J., Robinson, M.R., Slavin, J.A., Smith, D.E., Strom, R.G., Trombka, J.I., and Zuber, M.T. (2001) The MESSENGER mission to Mercury: scientific objectives and implementation. *Planetary and Space Science*, **49**, 1445–1465.

Sprague, A.L. (1990) A diffusion source for sodium and potassium in the atmospheres of Mercury and the Moon. *Icarus*, **84**, 93–105.

Sprague, A.L. (1992) Mercury's atmospheric sodium bright spots and potassium variations, a possible cause. *J. Geophys. Res.*, **97**, 18257–18264.

Sprague, A.L., and Roush, T.L. (1998) Comparison of laboratory emission spectra with mercury telescopic data. *Icarus*, **133**, 174–183.

Sprague, A.L., Kozlowski, R.W.H., and Hunten, D.M. (1990) Caloris Basin, an enhanced source for potassium in Mercury's atmosphere, *Science*, **249**, 1140–1143.

Sprague, A.L., Kozlowski, R.W.H., Witteborn, F.C., Cruikshank, D.P., and Wooden, D.H. (1994) Mercury, evidence for anorthosite and basalt from mid-infrared (7.5–13.5 μm) spectroscopy. *Icarus*, **109**, 156–167.

Sprague, A.L., Hunten, D.M., and Lodders, K. (1995) Sulfur at Mercury, elemental at the poles and sulfides in the regolith, *Icarus*, **118**, 211–215.

Sprague, A.L., Kozlowski, R.W.H., Hunten, D.M., Schneider, N.M., Domingue, D.L., Wells, W.K., Schmitt, W., and Fink, U., (1997) Distribution and abundance of sodium in Mercury's atmosphere, 1985–1988. *Icarus*, **128**, 506–527.

Sprague, A.L., Deutsch, L.K., Hora, J., Fazio, G.G., Ludwig, B., Emery, J., and Hoffmann, W.F. (2000) mid-Infrared (8.1–12.5 micrometer) imaging of mercury. *Icarus*, **147**, 421–432.

Sprague, A.L., Emery, J.P., Donaldson, K.L., Russell, R.W., Lynch, D.K., and Mazuk, A.L. (2002) Mercury, mid-infrared (3–13.5 micrometer) observations show heterogeneous composition, presence of intermediate and basic soil type, and pyroxene. *Meteoritics and Planetary Science*, **37**, 1255–1268.

Sprague, A.L., Nash, D.B., Witteborn, F.C., and Cruikshank, D.P. (1997) Mercury's feldspar connection, mid-ir measurements suggest plagioclase. *Adv. Space Res.*, **19**, 1507–1510.

Sprague, A.L., Schmitt, W.J., and Hill, R.E. (1998) Mercury, sodium atmospheric enhancements, radar bright spots, and visible surface features. *Icarus*, **135**, 60–68.

Sprague, A.L., and Roush, T.L. (1998) Comparison of laboratory emission spectra with mercury telescopic data. *Icarus*, **133**, 174–183.

Spudis, P., and Guest, J. (1988) Stratigraphy and geologic history of Mercury, In: F. Vilas, C. Chapman, and M. Matthews (eds), *Mercury*, pp. 118–164. Univ. of Arizona Press, Tucson, Arizona.

Stevenson, A. (1976) Crustal remanence and the magnetic moment of Mercury. *Earth Planet Sci. Lt.*, **28**, 454.

Strom, R.G. (1978) Origin and relative age of Lunar and Mercurian intercrater plains. *Phys. Earth Planet. Interiors*, **15**, 146.

Strom, R.G. (1979) Mercury: a post-Mariner 10 assessment. *Space Science Reviews*, **24**, 3–70.

Strom, R.G., and Knapp, W. (1984) Der Merkur [Mercury]. *Bild der Wissenschaft* [World of Science], Deutsche Verlags-Anstalt GmbH, Stuttgart, Germany.

Strom, R.G., (1984) Mercury. *Geology of the Terrestrial Planets*, NASA SP-469. Scientific and Technical Information Branch, National Aeronautics and Space Adinistration, Washington, DC..

Strom, R.G. (1987) *Mercury: The Elusive Planet*, solar system series. Smithsonian Institution Press, Washington, D.C.

Strom, R.G. (1990) Mercury. *McGraw-Hill Yearbook of Science and Technology*. McGraw-Hill Book Co, New York.

Strom, R.G. (1990) Mercury: Geology and geophysics. *Reference Encyclopedia of Astronomy and Astrophysics*. Van Nostrand Reinhold Book Co., New York.

Strom, R.G. (1990) Mercury: The forgotten planet. *Sky and Telescope Magazine*, **80**(3).

Strom, R.G. (1997) Mercury: An overview. *Advances in Space Research*, **19**(10), 1471–1485.

Strom, R.G. (1998) Mercury. *Encyclopedia of the Solar System*. Academic Press, New York.

Strom, R.G. (2000) Mercury. *Encyclopedia of Astronomy and Astrophysics*. Macmillian, Basingstoke, UK.

Strom, R.G., Malin, M.C., and Leake, M.A. (1990) *Geologic Map of the Bach Region of Mercury*, U.S. Geological Survey Map I-2015. US Geological Survey, Washington, DC.

Strom, R.G., and Neukum, G. (1988) The cratering record on Mercury and the origin of impacting objects, In: F. Vilas, C. Chapman, and M. Matthews (eds), *Mercury*, pp. 336–373. Univ. of Arizona Press, Tucson, Arizona.

Strom, R.G., *et al.* (1975) Preliminary imaging results from the second Mercury encounter, *J. Geophys. Res.*, **80**, 2345.

Strom, R.G., Trask, N.J., and Guest, J.E. (1975) Tectonism and volcanism on Mercury. *J. Geophys. Res.*, **80**, 2478.

Starukhina, L.V., and Shkuratov, Y.G. (2000) The lunar poles: water ice or chemically trapped hydrogen? *Icarus*, **147**, 585–587.

Taylor, L.A., Pieters, C.M., Morris, R.V., Keller, L.P., and McKay, D.S. (2001) Lunar mare soils, space weathering and the major effects of surface-correlated nanophase Fe. *J. Geophysical Res.*, **106**, 27985–28000.

Trask, N.J., and Guest, J.E. (1975) Preliminary geologic terrain map of Mercury. *J. Geophys. Res.*, **80**, 2461–2477.

Trask, N.J., and Strom, R.G. (1976) Additional evidence of Mercurian volcanism. *Icarus*, **28**, 559.

Tyler, A.L., Kozlowski, R.W.H., and Lebofsky, L.A. (1988) Determination of rock type on Mercury and the Moon through remote sensing in the thermal infrared. *Geophys. Res. Lett.*, **15**, 808–811.

Vasavada, A.R., Paige, D.A., and Wood, S.E. (1999) Near-surface temperatures on Mercury and the Moon and the stability of polar ice deposits. *Icarus*, **141**, 179–193.

Veverka, J., Helfenstein, P., Hapke, B., and Goguen, J.D. (1988) Photometry and polarimetry of Mercury, In: F. Vilas, C.R. Chapman, and M.S. Matthews (eds), *Mercury*, pp. 37–58. Univ. of Arizona Press, Tucson, Arizona.

Vilas, F. (1988) Surface composition of Mercury from reflectance spectrophotometry, In: F. Vilas, C.R. Chapman, and M.S. Matthews (eds), *Mercury*, pp. 59–76. Univ. of Arizona Press, Tucson, Arizona.

Vilas, F., Leake, M.A., and Mendell, W.W. (1984) The dependence of reflectance spectra of Mercury on surface terrain. *Icarus*, **59**, 60–68.

Warell, J., 2002. Properties of the Hermean regolith. II. disk-resolved multicolor photometry and color variations of the "unknown" hemisphere. *Icarus*, **156**, 303–317.

Warell, J. (2002b) Properties of the Hermean regolith. III. disk-resolved vis-NIR reflection spectra and implications for the abundance of iron. *Icarus*, **161**, 199–222.

Warell, J., and Lamaye, S.S. (2001) Properties of the Hermean regolith, I. global regolith albedo variation at 200 km scale from multicolor CCD imaging. *Planetary and Space Science Special Issue Mercury*, **49**, 1531–1552.

Watters, T.R., Schultz, R.A., and Robinson, M.S. (2000) Displacement-length relations of thrust faults associated with lobate scarps on Mercury and Mars: comparison with terrestrial faults. *Geophys. Res. Lt.*, **27**(22), 3659–3662.

Watters, T.R., Schultz, R.A., Robinson, M.S., and Cook, A.C. (2002) The mechanical and thermal structure of Mercury's lithosphere. *Geophys. Res. Letters*, **29**(11), 10.1029/2001.

Weidenschilling, S.J. (1978) Iron/silicate fractionation and the origin of Mercury. *Icarus*, **35**, 99–111.

Wetherill, G.W. (1988) Accumulation of Mercury from planetesimals, In: F. Vilas, C. Chapman, and M. Matthews (eds), *Mercury*, pp. 670–691. Univ. of Arizona Press, Tucson, Arizona.

Wilhelms, D.E. (1976) Mercurian volcanism questioned. *Icarus*, **28**, 551.

Index

Accretion
  equilibrium condensation model, 173
  fragmentation, 179
  models, 172–174
Adsorption and surface interaction, *see*
        Atmosphere, sinks
Albedo
  EUV, 92
  lunar highlands, 92
  visual, 91–92, 150
  UV, 92
ALPO (Amateur Lunar and Planetary
        Observers), 8–9, 150
Antipodal region, 125
Antoniadi, E., 6
Aphelion, 39–40
Arcuate scarps, *see* Lobate scarps and
        Faulting
Arecibo radar facility, 83
Astronomical symbol, 3
Atmosphere (exosphere)
  distribution, 69
  escape, 70
  known gases, 65–66, 68, 72
  Mariner 10, 27
  physics of, 66–69
  pressure, 67
  sinks, 70–72

sources, 70–72
velocities and temperatures, 69

Basalt
  composition of, 106
  flood, 152
  evidence for, 104
Ballistic trajectory, 118
Basins, 119–125
  Bach, 114
  Caloris, *see* Caloris Basin
Bepi Colombo mission, 184
Bow shock, 55–56

Ca *see* Calcium
Calcium
  atmospheric, 66, 68–70
  surface, 100–105
Calendar day on Mercury, 37–39, 149
Caloris basin, 76, 107, 121–125
  floor structure, 123–125, 128
  hummocky plains, 123
  mountains, 121
  ridges, 123
  sodium and potassium in association, 71
Charged particle experiment
  Mariner 10, 19
  MESSENGER, 183